I0053264

Optical Parametric Generation
and Amplification

Laser Science and Technology
An International Handbook

Editors in Chief

V.S. LETOKHOV, *Institute of Spectroscopy, Russian Academy of Sciences, 142092 Moscow Region, Troitsk, Russia*

C.V. SHANK, *Director, Lawrence Berkeley Laboratory, University of California, Berkeley, California 94720, USA*

Y.R. SHEN, *Department of Physics, University of California, Berkeley, California 94720, USA*

H. WALTHER, *Max-Planck-Institut für Quantenoptik und Sektion Physik, Universität München, D-8046 Garching, Germany*

This book is part of a series. The publisher will accept continuation orders which may be cancelled at any time and which provide for automatic billing and shipping of each title in the series upon publication. Please write for details.

Optical Parametric Generation and Amplification

Jing-yuan Zhang

Department of Physics, Georgia Southern University, USA

Jung Y. Huang

Institute of Electro-Optical Engineering, Chiao Tung University, Taiwan

and

Y.R. Shen

Department of Physics, University of California, USA, and
Materials Science Division, Lawrence Berkeley Laboratory, USA

CRC Press

Taylor & Francis Group

Boca Raton London New York

CRC Press is an imprint of the
Taylor & Francis Group, an **informa** business

CRC Press
Taylor & Francis Group
6000 Broken Sound Parkway NW, Suite 300
Boca Raton, FL 33487-2742

© 1995 by Taylor & Francis Group, LLC
CRC Press is an imprint of Taylor & Francis Group, an Informa business

No claim to original U.S. Government works

This book contains information obtained from authentic and highly regarded sources. Reason-able efforts have been made to publish reliable data and information, but the author and publisher cannot assume responsibility for the validity of all materials or the consequences of their use. The authors and publishers have attempted to trace the copyright holders of all material reproduced in this publication and apologize to copyright holders if permission to publish in this form has not been obtained. If any copyright material has not been acknowledged please write and let us know so we may rectify in any future reprint.

Except as permitted under U.S. Copyright Law, no part of this book may be reprinted, reproduced, transmitted, or utilized in any form by any electronic, mechanical, or other means, now known or hereafter invented, including photocopying, microfilming, and recording, or in any information storage or retrieval system, without written permission from the publishers.

For permission to photocopy or use material electronically from this work, please access www.copyright.com (http://www.copyright.com/) or contact the Copyright Clearance Center, Inc. (CCC), 222 Rosewood Drive, Danvers, MA 01923, 978-750-8400. CCC is a not-for-profit organiza-tion that provides licenses and registration for a variety of users. For organizations that have been granted a photocopy license by the CCC, a separate system of payment has been arranged.

Trademark Notice: Product or corporate names may be trademarks or registered trademarks, and are used only for identification and explanation without intent to infringe.

Visit the Taylor & Francis Web site at
http://www.taylorandfrancis.com

and the CRC Press Web site at
http://www.crcpress.com

CONTENTS

Introduction to the Series

Almost 30 years have passed since the laser was invented; nevertheless, the fields of lasers and laser applications are far from being exhausted. On the contrary, during the last few years they have been developing faster than ever. In particular, various laser systems have reached a state of maturity such that more and more applications are seen suffusing fields of science and technology, ranging from fundamental physics to materials processing and medicine. The rapid development and large variety of these applications call for quick and concise information on the latest achievements; this is especially important for the rapidly growing inter-disciplinary areas.

The aim of *Laser Science and Technology – An International Handbook* is to provide information quickly on current as well as promising developments in lasers. It consists of a series of self-contained tracts and handbooks pertinent to laser science and technology. Each tract starts with a basic introduction and goes as far as the most advanced results. Each should be useful to researchers looking for concise information about a particular endeavor, to engineers who would like to understand the basic facts of the laser applications in their respective occupations, and finally to graduate students seeking an introduction into the field they are preparing to engage in.

When a sufficient number of tracts devoted to a specific field have been published, authors will update and cross-reference their pages for publication as a volume of the handbook.

All the authors and section editors are outstanding scientists who have done pioneering work in their particular field.

V.S. Letokhov
C.V. Shank
Y.R. Shen
H. Walther

1. INTRODUCTION

High power, picosecond (or subpicosecond) coherent optical pulses, tunable over a wide spectral range from the ultraviolet (UV) to the infrared (IR), are most desirable in many applications.[1] They can be used, for example, in time-resolved spectroscopy to yield new information about fundamental properties of materials, to identify transient species, to characterize new nonlinear optical materials,[2] and to study the dynamics of optoelectronic systems.[3] In the past several decades, mode-locked dye[4] and solid state lasers[5] have been the major sources to provide picosecond as well as subpicosecond pulses in the visible and near infrared region with limited tuning ranges. Tunable mid-IR laser pulses are more difficult to generate, presumably because of the lack of suitable laser media. It is well known that nonlinear optical effects can be employed for frequency conversion[6] and that both second-order and third-order nonlinear optical processes are commonly used.[7] While second-order processes require a medium without inversion symmetry, the third-order processes can occur in any medium including gases and liquids. Particularly notable is frequency conversion by stimulated Raman scattering in molecular gases or atomic vapor. Hydrogen and methane are widely used as Raman shifters.[8] Stimulated electronic Raman scattering in alkali vapor has been employed to generate tunable mid-IR radiation.[9] Several drawbacks of such techniques should however be noted. First, the tuning range is often limited. Second, the conversion efficiency tends to be low and could fluctuate strongly. Finally, it is difficult to make the system compact.

Second-order nonlinear optical processes, such as sum-frequency generation (SFG), difference-frequency generation (DFG), and optical parametric oscillations[10] (OPO) and generation (OPG), are more commonly adopted for frequency conversion. They can be highly efficient and the systems are simple and compact. DFG and OPG are particularly attractive because they can yield an output with a very large tuning range extending from the visible to the infrared, limited only by absorption and phase matching of waves in the nonlinear crystal employed.

OPG and DFG are both wave-mixing processes involving energy conversion from a pump beam at frequency ω_3 into a signal beam at ω_1 and an idler beam at $\omega_2 = \omega_3 - \omega_1$. No clear distinction exists between the two. Usually, for DFG, one refers to a process with two intense input laser beams at ω_3 and ω_1, respectively, generating an output beam at the difference frequency $\omega_2 = \omega_3 - \omega_1$. For OPG, only a single laser beam at ω_3 is used as the input, and coherent outputs at ω_1 and ω_2 are generated. Often, one also speaks of optical parametric amplification (OPA), which is really not different from DFG except that one has in mind that the input at ω_1 is weak and is to be strongly amplified.

In this volume, we shall focus our discussion on OPG/OPA. The theory of OPG, OPA, or DFG was worked out by Armstrong, Bloembergen, Ducuing and Pershan early in 1962.[11] Giordinaine and Miller[12] first demonstrated the operation of an optical parametric oscillator. Wang and Racette[13] were the first to observe OPA in ammonium dihydrogen phosphate and Boyd *et al.*[14] succeeded in showing CW OPA in a lithium niobate crystal. Baumgartner and Byer,[15] later in 1979, conducted a detailed measurement of optical parametric gain in deuterated potassium dihydrogen phosphate and lithium niobate and compared the results with theory. More recently, Kaiser, Piskarskas and their co-workers, as well as others, have contributed significantly to the progress of the field.[1] However, because of difficulties originating from poor laser beam quality and unavailability of suitable nonlinear optical crystals, OPG/OPA as a viable, coherent, tunable light source was not duly recognized until recently. The situation has changed following the recent advancement in laser technology and nonlinear optical crystals. Highly stable pulsed lasers with good beam quality are now available.[16] Nonlinear crystals with wide tuning range and high laser damage thresholds have been developed.[17] It then becomes possible to construct a suitable OPG/OPA system for routine measurements in a laboratory. Indeed, high-energy, widely tunable, picosecond as well as subpicosecond pulsed OPG/OPA has recently attracted much attention. Picosecond and subpicosecond output pulses with an energy of ~ 1 mJ/pulse at a conversion efficiency as high as 30% and a tuning range more than 20,000 cm^{-1} can be generated. To cover the same tuning range by a dye laser would require at least 20 different dyes, not counting the infrared range where laser dyes are presently not available.

In general, tunable optical parametric devices can be divided into two groups: OPO and OPG/OPA. An OPO is composed of a nonlinear crystal situated in a resonant cavity. The case of pumping by nanosecond pulses has been analyzed in detail by Bronsnan and Byer[18] and the case of synchronous pumping by picosecond or femtosecond mode-locked pulses is described in a concurrent volume by Tang and Cheng.[19] An OPG/OPA device, on the other hand, involves no cavity and the tunable output is generated from noise (or parametric fluorescence) and amplified in the traveling wave form.[20] This latter case is what we would like to concentrate on in this volume.

An OPG/OPA system can have many advantages. It is simple and straightforward in construction. Without a cavity, the output tuning range is only limited by phase-matching and transparency of the nonlinear crystals. High energy (or intensity), widely tunable, picosecond or subpicosecond pulses are readily obtainable. Because of the absence of a resonant cavity, the single-pass gain of OPG/OPA must be high in order to yield a strong output. This calls for a pump of high intensity. In practical cases, the pump intensity needed is in the GW/cm^2 range or higher; so high-energy picosecond or subpicosecond pump lasers must be employed.

This volume is organized as follows: We first outline in Section 2 the theoretical background for design of an OPG/OPA system, using two newly discovered nonlinear crystals, β-barium borate (BBO) and lithium triborate (LBO) as examples. We then describe in Section 3 experimental design considerations, including the selection of the nonlinear crystals, pumping sources, and optical configurations. Section 4 reviews the experimental results and their comparison with theoretical calculations. We also discuss the scheme used to narrow the spectral width of the OPA output, and the scheme that can extend the tuning range of an OPG/OPA system from near IR to mid-IR by means of DFG and from visible to UV using SFG.

2. THEORETICAL BACKGROUND

The theory of optical parametric processes in a nonlinear medium follows the earlier work of Armstrong et al.[11] and has been reviewed by many authors.[7,15] Here, we focus only on the essential points.

2.1 OPTICAL PARAMETRIC AMPLIFICATION AS A THREE-WAVE MIXING PROCESS

Optical parametric amplification (OPA) is a three-wave mixing process, in which a pump beam, a signal beam, and an idler beam at frequencies ω_3, ω_1, and ω_2, respectively, with $\omega_3 = \omega_1 + \omega_2$, propagate and interact in a nonlinear medium. We consider here the collinear propagation geometry. The three waves propagating along $\hat{z}$ are coupled through the second-order nonlinear polarization $\vec{P}_{NL}(r, t)$ and are governed by the wave equation:

$$\nabla \times \nabla \times \vec{E}(r, t) + \frac{1}{c^2}\frac{\partial^2}{\partial t^2}\vec{D}(r, t) = -\frac{4\pi}{c^2}\frac{\partial^2}{\partial t^2}\vec{P}_{NL}(r, t) \qquad (1)$$

Here, $\vec{E}(r, t)$ and $\vec{D}(r, t)$ are the total electric field and displacement current given by

$$\vec{E}(r, t) = \sum_{n=1}^{3}\left[\hat{e}_n\vec{A}_n(r, t)e^{i(\vec{k}_n\cdot\vec{r}-\omega_n t)} + c.c.\right] \qquad (2)$$

and $\vec{D}(r, t) = \vec{E}(r, t) + 4\pi\vec{P}_L(r, t)$, and $\vec{A}_n$ is the complex slowly varying amplitude of the electric field of the n-th wave. The linear and nonlinear polarizations $\vec{P}_L$ and $\vec{P}_{NL}$, are related to the field by the constitutive relations[21]

$$\vec{P}_L(r, t) = \int_{-\infty}^{+\infty} \vec{\chi}^{(1)}(t - t') \cdot \vec{E}(r, t')dt'$$

$$\vec{P}_{NL}(r, t) = \int_{-\infty}^{+\infty}\int_{-\infty}^{+\infty} \vec{\chi}^{(2)}(r, t - t_1, t - t_2) : \vec{E}(r, t_1)\vec{E}(r, t_2)dt_1 dt_2 \qquad (3)$$

which are valid in the electric-dipole approximation. If the three waves are pulsed and quasi-monochromatic, $\chi^1(t - t')$ can be expanded into a Taylor series of $(t' - t)$ to reflect the group-velocity mismatch and pulse spreading caused by material dispersion.

Considerable simplification occurs if the second-order response is assumed to be instantaneous. We then have $\vec{P}_{NL}(r, t) = (\vec{\chi}^{(2)} : \vec{E}(r, t)\vec{E}(r, t)$. The assumption amounts to neglecting the dispersion of $\vec{\chi}^{(2)}$. It is justified for pulse widths >10 fs assuming the electronic contribution to $\vec{\chi}^{(2)}$ dominates.

2.2 COUPLED WAVES

2.2.1 Nonlinear polarization

With the assumption of negligible dispersion in $\vec{\chi}^{(2)}$, the nonlinear polarization in Eq. (3) can be decomposed into different frequency components. Those at frequencies ω_1, ω_2 and ω_3 are relevant to the optical parametric process. They are

$$P_{NL}(\omega_1) = \chi_{\text{eff}}^{(2)} A_3 A_2^* e^{i(\vec{k}_3 - \vec{k}_2) \cdot \vec{r} - i\omega_1 t}$$

$$P_{NL}(\omega_2) = \chi_{\text{eff}}^{(2)} A_3 A_1^* e^{i(\vec{k}_3 - \vec{k}_1) \cdot \vec{r} - i\omega_2 t}$$

$$P_{NL}(\omega_3) = \chi_{\text{eff}}^{(2)} A_1 A_2^* e^{i(\vec{k}_3 + \vec{k}_2) \cdot \vec{r} - i\omega_3 t}$$

where $\chi_{\text{eff}}^{(2)}$ is the effective nonlinearity which depends on the beam polarization and the symmetry property of $\vec{\chi}^{(2)}$ tensor.[7,11]

2.2.2 Coupled-wave equations

In the slowly varying amplitude approximation (SVA), Eq. (1) can be simplified and decomposed into three coupled-wave equations

$$\hat{D}_1 A_1 = i(2\pi\omega_1/cn_1)\chi_{\text{eff}}^{(2)} A_3 A_2^* e^{i\Delta kz}$$

$$\hat{D}_2 A_2 = i(2\pi\omega_2/cn_2)\chi_{\text{eff}}^{(2)} A_3 A_1^* e^{i\Delta kz} \tag{4}$$

$$\hat{D}_3 A_3 = i(2\pi\omega_2/cn_3)\chi_{\text{eff}}^{(2)} A_1 A_2^* e^{i\Delta kz}$$

where $\Delta k = k_3 - k_1 - k_2$ indicates the phase mismatch and $\hat{D}_n$ are the differential operators defined by[22]

$$\hat{D}_n = \frac{\partial}{\partial z} + \rho_n \frac{\partial}{\partial x} + \frac{i}{2k_n}\left(\frac{\partial^2}{\partial x^2} + \frac{\partial^2}{\partial y^2}\right) + \frac{1}{v_n}\frac{\partial}{\partial t} + ig_n\frac{\partial}{\partial t^2} \tag{5}$$

where ρ_n is the walk-off angle between the wave and ray vectors, $v_n = c/n_{gn}$ with $n_{gn} = n_n + \omega_n(\partial n_n/\partial \omega)|_{\omega_n}$ is the group velocity, and $g_n = 1/2(\partial^2 k/\partial \omega^2)|_{\omega_n}$ is the dispersion-spreading coefficient.

2.2.3 Effective lengths of optical parametric interaction

Within the stated approximations, the solution of Eq. (4) describes OPA, including beam diffraction and dispersion, in an anisotropic crystal. In general, Eq. (4) cannot be solved analytically so that numerical solution is needed. In many cases, Eq. (4) can be further simplified with more approximations.

To facilitate discussion of additional approximations, let us first define a number of characteristic lengths for the OPA process:[23]

(a) Aperture length, $L_{an} = d_o/\rho_n$, for a beam with beam diameter d_o and walk-off angle ρ_n between the signal (or idler) and pump beams: this describes the propagation distance over which the extraordinary beam is transversely displaced by d_o; (b) Quasi-static interaction length, $L_{qs} = \tau(1/v_1 - 1/v_2)^{-1}$, for two pulses with group velocities v_1 and v_2 and pulse width τ: this is the distance over which the two pulses experience a relative time delay τ; (c) Diffraction length, $L_{\text{dif}} = kd_o^2$: this denotes the distance over which the beam diameter increases by a factor of $2^{1/2}$ due to diffraction; (d) Dispersion-spreading length, $L_{ds} = \tau^2/g_n$: this indicates the propagation distance the pulse would take to double its pulsewidth because of dispersion; and (e) Nonlinear interaction length defined as $L_{nl} = cn/[2/\pi\omega A_o\chi_{\text{eff}}^2]$: this is the length of the nonlinear medium required to have significant energy transfer among the coupled waves assuming that they are infinite plane waves. For example, as can be seen from Eq. (7) and Eq. (8) in the subsection that in the case of SHG, the conversion efficiency $\eta = 58\%$ when $L_c = L_{nl}$; these five characteristic lengths have separate bearings on the four terms in the differential operator $\vec{D}_n$ in Eq. (5) and one term from the inhomogeneous part of Eq. (5). It can be seen physically that if the length of the medium is much shorter than a particular characteristic length, then the term in $\vec{D}_n$ corresponding to this characteristic length can be neglected. The coupled-wave equations can thus be simplified.

2.2.4 Solution with a depletionless pump beam

The simplest case occurs when the nonlinear crystal has a length L_c which is much shorter than all the above characteristic lengths except L_{nl}. Equation (4) then becomes

$$\frac{\partial A_1}{\partial z} = i \frac{2\pi\omega_1}{n_1 c} \chi_{\text{eff}}^{(2)} A_3 A_2^* e^{i\Delta kz}$$

$$\frac{\partial A_2}{\partial z} = i \frac{2\pi\omega_2}{n_2 c} \chi_{\text{eff}}^{(2)} A_3 A_1^* e^{i\Delta kz} \tag{6}$$

$$\frac{\partial A_3}{\partial z} = i \frac{2\pi\omega_3}{n_3 c} \chi_{\text{eff}}^{(2)} A_1 A_2^* e^{i\Delta kz}$$

In the small-signal amplification limit (i.e., $L_c < L_{nl}$), the pump depletion is negligible, so that A_3 can be regarded as a constant $[I_3(z) = I_3(0)]$. The output intensities $I_i = \frac{cn_i}{2\pi}|A_i|^2$ of the signal and idler waves from OPA, calculated from Eq. (6) with an initial condition $I_3(0) = I_{30} \gg I_{10}$, and $I_{20} = 0$, are given by[15]

$$I_1(z) = I_{10} \cosh^2(gz)$$

$$I_2(z) = I_{10} \frac{\omega_2}{\omega_1} \sinh^2(gz) \tag{7}$$

$$I_3(z) = I_{30}$$

where $g = \left[g_0^2 - \frac{\Delta k^2}{4} \right]^{1/2}$, with $g_0 = \left(\frac{8\pi^3}{c^3} \frac{\omega_1\omega_2}{n_1 n_2 n_3} \chi_{\text{eff}}^{(2)^2} I_{30} \right)^{1/2}$ being the steady-state small signal gain. The maximum gain occurs at phase matching, $\Delta k = 0$, which governs the tuning curve of an optical parametric device and will be discussed in detail in Section 2.3.

2.2.5 Solution with pump depletion

If $L_c > L_{nl}$ in the above case, strong energy exchange is expected among the three coupled waves. Depletion and later recovery of the pump beam can occur in the process. The solution of Eq. (6) with $\Delta k = 0$ and $I_{20} = 0$ can be expressed in terms of the Jacobian elliptic function, sn:[15]

$$I_1(z) = I_{10} + \frac{\omega_1}{\omega_3} I_{30} \left\{ 1 - sn^2 \left[Ng_0z - \frac{K(N)}{N^{1/2}}, N \right] \right\}$$

$$I_2(z) = \frac{\omega_2}{\omega_3} I_{30} \left\{ 1 - sn^2 \left[Ng_0z - \frac{K(N)}{N^{1/2}}, N \right] \right\} \tag{8}$$

$$I_3(z) = I_{30} \, sn^2 \left[Ng_0z - \frac{K(N)}{N^{1/2}}, N \right]$$

where $K(N)$ is the complete elliptical function of the first kind and $N = \frac{1}{1+(\omega_3 I_{10})/(\omega_1 I_{30})}$ is the ratio of the number of photons in the input pump beam to the total input photon number, or the normalized photon number.

2.2.6 Effects of the spatial and temporal profiles of the coupled beams

The spatial and temporal profiles of a pump pulse can have profound effect on the conversion efficiency of an optical parametric device. They can lead to significant deviations from the prediction of Eq. (8) if $L_c \geq L_{\text{dif}}$ and $L_c > L_{qs}$. On the other hand, if $L_c \ll L_{\text{dif}}, L_{qs}$, then the spatial and temporal profiles of the pump beam can be approximately taken into account by subdividing them into histograms.[23,24] For each step in the histograms, the output is calculated with Eq. (8). The total output is obtained by summing up contributions from all the steps. This procedure will be used in the first-order design of OPG/OPA discussed later in Sections 2.4 and 2.5.

Historically, the conversion efficiency of quasi-static OPA of focused Gaussian beams $[(L_a, L_{\text{dif}}) < L_c \ll (L_{qs}, L_{ds}, L_{nl}]$ has been investigated theoretically by Boyd and Kleinman[25] in the pump depletionless limit. For the case of appreciable energy exchange between waves, a split-step Fourier transform technique[26] has been developed. It is based on the assumption that in propagating the waves over a small distance, the diffraction and nonlinear interaction of waves can be treated independently. The method is implemented by first dividing the crystal into a large number of segments of thickness h. The optical waves at position z, assumed to be non-diffracted, are allowed to propagate and interact nonlinearly for a distance of $h/2$ over which the Runge-Kutta procedure is used to solve the coupled-wave equations. At $z + h/2$, the field is Fourier transformed and multiplied by a term describing the effect of diffraction over the whole segment length h. Finally, the field is recovered by inverse Fourier transform and propagated the remaining distance of $h/2$ with nonlinear interaction to obtain the optical field for propagation to the next segment. The fast Fourier transform (FFT) algorithm can be used to make the numerical calculation relatively fast.

More generally, Eq. (4) should be solved directly in order to take into account the effect of the spatial and temporal profiles of the beams. The most widely used method for numerical solution of Eq. (4) is the finite-difference method[27] which is usually slower by up to two orders of magnitude in comparison with the efficient split-step scheme. Moreover, considerable computational difficulties have been encountered in the finite-difference method.

2.3 WAVELENGTH TUNING IN OPTICAL PARAMETRIC DEVICES

The output frequencies of a parametric device are governed by energy ($\omega_3 = \omega_1 + \omega_2$) and momentum conservation (or phase matching, $\Delta k = k_3 - k_1 - k_2 = 0$).[7] They can be combined to form

$$\omega_3 n_3(\omega_3) = \omega_1 n_1(\omega_1) = \omega_2 n_2(\omega_2) \qquad (9)$$

The equation can be satisfied in an anisotropic crystal with birefringence compensating normal dispersion. This can be seen clearly if Eq. (9) is written as

$$\omega_3 = [n_{3e}(\omega_3, \theta) - n_{3o}(\omega_3)] = \omega_1 n_{1o}(\omega_1) + \omega_2 n_{2o}(\omega_2) - \omega_3 n_{3o}(\omega_3) \qquad (10a)$$

for a negative uniaxial crystal, and

$$\omega_3 [n_{3o}(\omega_3) - n_{3e}(\omega_3, \theta)] = \omega_1 n_{1e}(\omega_1) + \omega_2 n_{2e}(\omega_2) - \omega_3 n_{3e}(\omega_3, \theta) \qquad (10b)$$

for a positive uniaxial crystal. Here, the subindices o and e refer to ordinary and extraordinary waves. If the beam at frequencies ω_1 and ω_2 are both ordinary or extraordinary, we have the so-called type-I phase matching case. If one of them is ordinary and the other extraordinary, we have a type-II phase matching case. Fig. 2.1 shows the refractive indices as functions of wavelength for ordinary and extraordinary waves in negative and positive crystals.

Equation (9) or (10) indicates that the output frequencies, ω_1 and ω_2, can be tuned if $n(\omega)$ is varied by an external parameter. This can be done by changing either the orientation (angle tuning) or temperature (temperature tuning) of the crystal.

In considering, for example, the tuning of a negative uniaxial crystal in the type-I phase matching case, we have $n_e^{-2}(\omega, \theta) = \frac{\sin^2 \theta}{n_E^2(\omega)} + \frac{\cos^2 \theta}{n_o^2(\omega)}$, where θ is the angle between the crystal c-axis and the beam propagation direction, and $n_E(\omega) \equiv n_e(\omega, \theta = 0) \leq n_e(\omega, \theta)$. Eq. (9) can be written as

$$\Delta n = +[n_0(\omega_3) - n_e(\omega_3, \theta] = 0 \qquad (11)$$

where $\Delta n = n_E(\omega_3) - n_o(\omega_1)\frac{\omega_1}{\omega_3} - n_o(\omega_2)\frac{\omega_2}{\omega_3}$. For a given θ, Eq. (11) can be satisfied by a certain set of ω_1 and ω_2. If the crystal is rotated to $\theta + \Delta\theta$, the output frequencies change accordingly to $\omega_1 + \Delta\omega$ and $\omega_2 - \Delta\omega$. Given the dispersions of $n_o(\omega)$ and $n_e(\omega, \theta)$, the angle tuning curve of ω_1 (or ω_2) versus θ can be calculated from Eq. (11).

Temperature tuning is particularly attractive if phase matching can be achieved with the crystal orientation at $\theta = 0$, known as noncritical phase matching (NCPM).

$n(\omega)$ (a. u.)

(a) type-I: o + o --- e

$n_o(\omega)$

$n_e(\omega, \theta)$

$n_E(\omega)$

ω 2ω

Frequency (a. u.)

a

$n(\omega)$ (a. u.)

(b) type-II: e + e --- o

$n_E(\omega)$

$n_e(\omega, \theta)$

$n_o(\omega)$

ω 2ω

Frequency (a. u.)

b

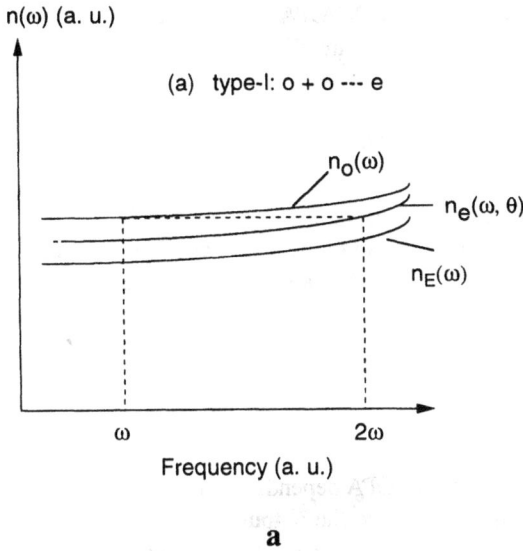

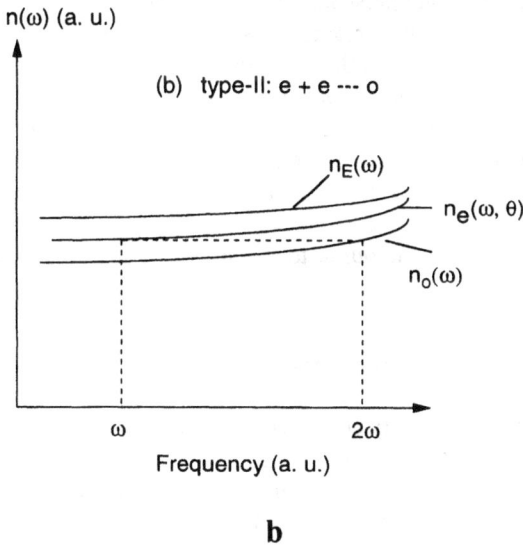

Fig. 2.1 Schematics showing the refractive indices as function of wavelength for ordinary $n_o(\omega)$ and extraordinary $n_e(\omega)$ in a (a) negative and (b) positive nonlinear crystal. Type-I and type-II phase matching for second harmonic generation is achieved in (a) and (b) respectively.

It has the advantage of having a much larger acceptance angle than the usual case. The phase matching condition OPG/OPA can be written in a form similar to Eq. (11) with angle θ replaced by temperature T.

2.4 ENERGY CONVERSION

Most important in the design of an efficient OPG/OPA system is maximization of energy conversion from the pump to the signal and idler beams. Several parameters are known to be important. To understand their roles in an optical parametric device, we have conducted a series of calculations using the numerical procedure described earlier.

2.4.1 Effect of the crystal length

To illustrate how the output of OPA depends on the crystal length and pump intensity, we have used Eq. (8) to calculate the output of a type-I BBO OPA pumped by the third-harmonic output of a picosecond active/passive mode-locked Nd:YAG laser with a fixed input signal of about 5 μJ at 550 nm and a near-Gaussian beam diameter of 2 mm.[28,29] The results are presented in Fig. 2.2. The output shows an oscillatory behavior as the crystal length increases. This is expected because reversal in energy conversion from the signal and idler back to the pump would occur if the former become sufficiently strong. Due to spatial and temporal variations of the pump pulse, energy conversion is expected to be different for different parts of the pulsed beams, so that the contrast of maxima and minima in the oscillation of the output in Fig. 2.2 is highly smeared. It is further reduced with increase of the crystal length as the spatial and temporal profiles of the beams evolve into more complex forms. The highest conversion efficiency from the pump to the signal occurs around the first peak in Fig. 2.2.

2.4.2 Effect of the pump intensity

The effects of the pump intensity and crystal length on OPA are usually complementary. The calculated results shown in Fig. 2.2 indicate that a short crystal pumped at a higher intensity appears to yield a better conversion efficiency. For example, for a fixed signal input of 30 μJ, a more efficient operation can be achieved with shorter crystals (<10 mm) and higher pump intensities (>3.5 GW/cm^2). The energy conversion efficiency from the pump to the signal for a 7 mm LBO crystal pumped at 5.8 GW/cm^2 can be about 30%.[30] In practice, however, the pump intensity is limited by the optical damage threshold of the material. For a reliable operation of OPA,

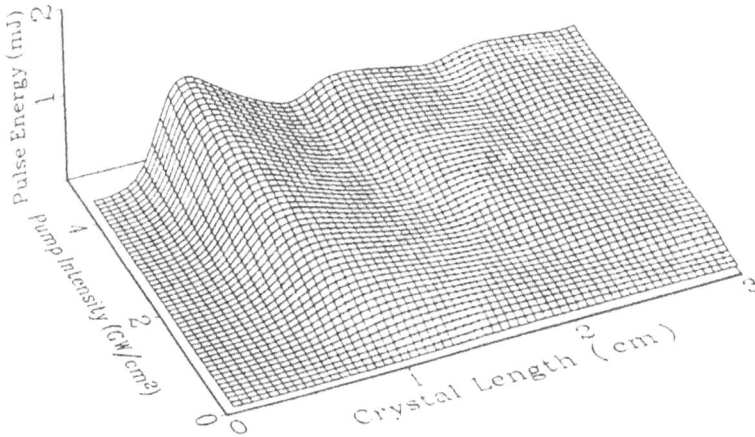

Fig. 2.2 Calculated output signal energy as a function of crystal length and pump intensity for an optical parametric amplifier composed of a type-I phase-matched β-BaB$_2$O$_4$ (BBO) crystal pumped at 355 nm. The seed signal energy is 5 μJ at a wavelength of 550 nm and a beam diameter of 0.2 cm.

a crystal length is often chosen such that OPA is operated at a pump intensity about a factor of 2–5 lower than the damage threshold.

2.4.3 Effect of the seed beam intensity

The injection signal is also an important parameter for optimizing the performance of an OPA. Fig. 2.3 shows the output signal as a function of the injected signal for a BBO OPA pumped at 355 nm.[29] The pump intensity in this calculation is 2.4 GW/cm^2 (2.4 mJ with a 2.8 mm beam spot). The results for three different signal wavelengths, which nearly span the whole tuning range of the signal beam, are shown by the solid (0.46 μm) and long-dashed (0.57 μm) and short-dashed (0.66 μm) curves respectively. It is interesting to note that they all have a broad region where the output signal appears to be insensitive to the injected energy. In this region, the OPA is expected to be more stable. With higher injection energy (>50 μJ), the energy transfer from the pump to the signal is expected to be more significant although no experimental result has yet been reported. It is also expected that the higher conversion efficiency is obtained at the expense of stability of the device.

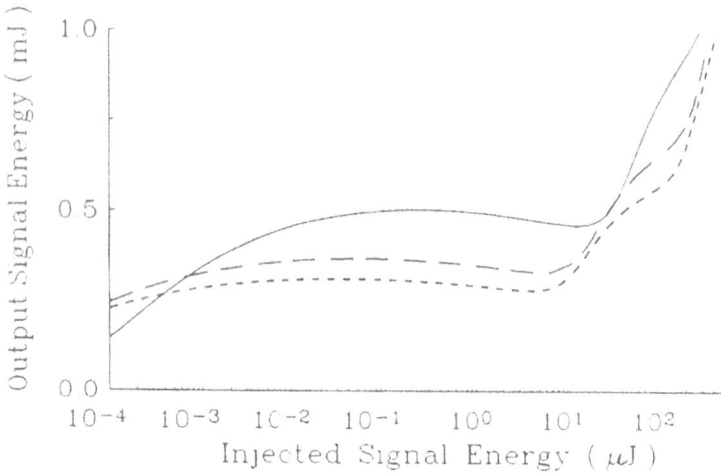

Fig. 2.3 Calculated signal output energy vs. injected energy for a BBO OPA. The pump energy is 2.4 mJ at 355 nm with a 15-ps pulse width, telescoped to a 2.8 mm beam spot. Solid, long-dashed and short-dashed curves refer to output wavelength at 460, 570 and 660 nm respectively. The crystal length is 1.5 cm.

2.5   OUTPUT LINEWIDTH

Since OPG is initiated from noise, the output bandwidth of an OPG/OPA system is generally quite broad. Several factors are responsible for the bandwidth of the signal and idler waves from an OPG/OPA system. For example, the bandwidth of the early OPG/OPA in LiNbO$_3$ by Laubereau *et al.*[20,40] was about 100 cm^{-1} if no special measure was taken to narrow down the bandwidth. Such a broad band radiation may have some limit in many applications. Great efforts have been made in narrowing the bandwidth since the appearance of first OPG/OPA system. Seilmeier *et al.*[31] first investigated the physical processes that contribute to the bandwidth of the signal and idler beam in a picosecond parametric device with high pump intensity. The factors that can affect the bandwidth of an OPG device are briefly discussed here.

2.5.1   Collinear phase mismatching

In a collinear phase matching geometry, the output of an OPG starting from noise is nearly proportional to exp$(2gL)$, if the pump depletion is negligible, where g is the gain coefficient with $g = \left[g_o^2 - \left(\frac{\Delta K}{2}\right)^2\right]^{1/2}$ with g_0 being the steady-state small signal gain as defined in Eq. (7), and L is the interaction length. The amplification is reduced by a factor of two when the phase mismatch Δk has the value of $\Delta k \approx 2[(\ln 2)g_o/L]^{1/2}$. This corresponds to a frequency deviation $\Delta \nu$

$$\Delta \nu = \frac{\Delta k}{2\pi (n_{\text{effi}} - n_{\text{effs}})} \qquad \text{for } \nu_s \neq \nu_i \qquad (12)$$

where $n_{\text{effs},i} = c/\nu_{s,i}$ is the effective refractive index of the nonlinear crystal and $\nu_{s,i}$ is the group velocity of the signal (or idler) wave in the crystal.

2.5.2 Non-collinear phase matching

The phase matching condition $\vec{k}_1 + \vec{k}_2 = \vec{k}_3$ can also be satisfied at other frequencies for wave mixing in noncollinear geometry. With $\Delta \vec{k} = 0$, the frequency shift $\Delta \nu = \nu_s(\alpha) - \nu_s(0)$ for a signal wave propagating off-axis with an external angle α is given by:

$$\Delta \nu = \alpha^2 \frac{\nu_p n_p \nu_s}{2(n_{\text{effs}} - n_{\text{effi}})\nu_i n_i n_s} \qquad (13)$$

The effective allowed angle α should be estimated from the geometry of the OPG system, or from the ratio of the beam diameter to the crystal length in a single crystal OPG.

2.5.3 Divergence of the pump beam

The divergence of the pump beam may also affect the bandwidth of an OPG/OPA system. Consider a pump beam with divergence angle ϕ (or ϕ/n_p inside the crystal). This leads directly to a frequency broadening of $\Delta \nu \sim (\phi/n_p)d\nu/d\theta$, where $d\nu/d\theta$ represents the slope of the tuning curve and n_p is the refractive index of the crystal seen by the pump beam. The value of $d\nu/d\theta$ depends on the phase matching angle θ. This broadening mechanism is most significant near the degenerate point. The beam divergence of a commercial active/passive mode-locked Nd:YAG laser system can be as small as 0.2 mrad; the corresponding bandwidth of the signal beam at 550 nm from a BBO-OPG/OPA pumped at 355 nm is estimated to be about 0.2 nm. Such a bandwidth is small compared with those resulting from the other broadening mechanism mentioned above and could be negligible. However, it becomes significant when a transform-limited bandwidth is desired, as will be discussed next. In the case of noncritical phase matching, $d\nu/d\theta = 0$ and the frequency broadening due to divergence of the pump beam is greatly reduced.

2.5.4 Bandwidth of the pump beam

The finite frequency width of the pump beam also affects the bandwidth of the signal (or idler) beam. The effect is complicated by many associated factors for femtosecond pump pulses. For longer transform-limited pump pulses, however, it is usually negligible since the phase matching condition changes insignificantly with the pump frequency within the bandwidth.

2.5.5 Power broadening

The output bandwidth can also be broadened by the pump power broadening. As mentioned in Section 2.5.1, the bandwidth due to collinear phase mismatch is proportional to $\Delta k \approx 2[(\ln 2)g_o/L]^{1/2}$ and, hence, proportional to $I_p^{1/4}$. This has not yet taken into account broadening due to saturation of amplification resulting from pump depletion. For an OPG/OPA system with a nearly transform-limited output bandwidth, power broadening could be the major mechanism that limits the output bandwidth.[32] More details will be discussed later.

2.5.6 Temperature broadening

In a temperature tuned OPG/OPA, the crystal parameters and the pump frequency are usually so chosen that the output frequency of OPG/OPA is very sensitive to the temperature of the crystal in order to have a broad tuning range within a limited temperature range. The temperature of the crystal is often controlled by a temperature-regulated oven. Since most nonlinear optical crystals are poor thermal conductors, temperature gradient may exist in the crystal and lead to appreciable output spectral broadening. Take LBO (type-I cut for NCPM and pumped at 532 nm) as an example. If the temperature changes from 110 to 145°C, the signal frequency of the parametric output changes from 17,700 cm^{-1} to 9,400 cm^{-1}, corresponding to 240 cm^{-1} per degree on average.[33] Clearly even a small temperature gradient in the crystal could result in a significant broadening in the output bandwidth.

3. EXPERIMENTAL CONSIDERATIONS

In designing a practical OPG/OPA device, the following steps are encountered: selection of the nonlinear crystal, selection of the pump source, and selection of configuration of the system.

3.1 MATERIAL CONSIDERATIONS

A nonlinear crystal is suitable for an OPG/OPA system if it possess the following qualities: (1) phase-matchable in the desired tuning range either by angle or temperature tuning; (2) possessing sufficiently large nonlinearity and high laser damage threshold; (3) having small absorption and scattering loss at both pump and signal frequencies; (4) chemically stable in the working environment; (5) large and homogeneous enough for efficient energy conversion over a wide tuning angle; (6) sufficiently small walk-off angle in order not to limit the interaction length; (7) small group velocity mismatch between the pump and the generated pulses especially for femtosecond OPG/OPA.

3.1.1 Materials for visible and near-ultraviolet output

Among the available nonlinear optical crystals, β-barium metaborate (β-BaB$_2$O$_4$ or BBO) and lithium triborate (LiB$_3$O$_5$ or LBO) are most suitable for OPG/OPA to generate tunable output in the visible and near UV. BBO is transparent from 2.6 μm to 191 nm, while LBO is transparent from 2.7 μm to 160 nm. Both BBO and LBO are mechanically hard and chemically stable. BBO is slightly hygroscopic while LBO is non-hygroscopic. The only problem with LBO is that, after exposure to air for several months, the surfaces may turn gray, which cannot be removed by surface cleaning. The mechanism of such a surface modification is still not understood. To avoid this problem, protective coating must be used, or the crystal must be kept either in dry N$_2$ or at an elevated temperature. Both BBO and LBO have very high nonlinearities with $\chi_{eff}^{(2)} = 6.7 \times 10^{-9}$ esu (or $d_{eff}^{(2)} = 1.4$ pm/V) for BBO[34] and $\chi_{eff}^{(2)} = 5.3 \times 10^{-9}$ esu (or $d_{eff}^2 = 1.1$ pm/V) for LBO[35] at the frequency degenerate point when pumped at 355 nm. They also have very high laser damage thresholds. With a 10 Hz active/passive mode-locked Nd:YAG laser as the pump source, the short-term damage threshold of BBO is about 20 GW/cm^2 for 1.064 μm and 35 ps pulses, and is about 15 GW/cm^2 for 532 nm and 20–25 ps, and 8–10 GW/cm^2 for 355 nm and 15 ps.[34] For LBO, it is more than 25 GW/cm^2 at 1.064 μm, about 15–20 GW/cM^2 at 532 nm and more than 12 GW/cm^2 at 355 nm. The pumping intensity for BBO that permits long-term operation without damage is around 7–8 GW/cm^2 at 532 nm and around 5 GW/cm^2

at 355 nm.[35] For LBO, the corresponding pumping intensity is about 10 GW/cm^2 at 532 nm and 7 GW/cm^2 at 355 nm, respectively. The values are based on the assumption that the pump beam has nearly a Gaussian TEM$_{00}$ mode.

KDP and KD*P crystals can also be used in a visible/UV OPG/OPA system. Both are transparent from 200 nm to 1.5 μm with a nonlinear coefficient about one third of LBO. The laser damage thresholds of KDP and KD*P are slightly lower than those of BBO and LBO. The advantage of using KDP or KD*P is that large-size crystals with high optical quality can be obtained at an affordable price. By selecting the pumping parameters properly, it is possible to construct an OPG/OPA system using KDP or KD*P with a fairly high conversion efficiency.[39,49]

3.1.2 Materials for visilble and infrared output

Potassium titanyl phosphate (KTP, KTiOPO$_4$) and lithium niobate (LiNbO$_3$) are both chemically stable and non-hygroscopic. KTP is transparent from 0.35 to 4.5 μm and LiNbO$_3$ is transparent from 0.4 to 5 μm. They are phase matchable for OPG/OPA in the spectral range of 0.6–4.0 μm. However, due to the orthophosphate overtone absorption at 3.5 μm in KTP, the output of KTP OPG/OPA drops significantly beyond 3.5 μm, as will be shown later. The laser damage threshold of LiNbO$_3$ is low, about or less than 10^9 W/cm^2 for picosecond pulses at 1.064 μm. The damage threshold of KTP is higher. For 30-picosecond laser pulses at 1.064 μm, it is about 3–5 GW/cm^2 for the flux grown crystals and 10–15 GW/cm^2 for the hydrothermally grown crystals. It drops significantly with pumping in the visible due to the presence of two-photon absorption. When pumped at 1.064 μm, the nonlinearities of both KTP with $\chi_{\text{eff}}^{(2)} = 1.49 \times 10^{-8}$ esu (or $d_{\text{eff}}^2 = 3.2$ pm/V)[36] and LiNbO$_3$ with $\chi_{\text{eff}}^{(2)} = 2.7 \times 10^{-8}$ esu (or $d_{\text{eff}}^2 = 5.8$ pm/V)[37] are much higher than that of BBO, and therefore even with the lower damage thresholds both crystals are still very attractive candidates for OPG/OPA in the near IR, especially in the range from 2.5 to 3.5 μm, where both BBO and LBO cannot function. The damage threshold of LiNbO$_3$ can be greatly improved by MgO doping.[38] Unfortunately, the doped crystals appear to be less homogeneous and not of sufficiently high quality to be suitable for OPA.

Two other crystals suitable for OPG/OPA in the near- to mid-infrared are potassium niobate (KNbO$_3$) and potassium titanyl arsenate (KTA, KTiOAsO$_4$). KNbO$_3$ is transparent from 400 nm down to 5.5 μm and has a very high nonlinear coefficient of 16–20 pm/V when pumped at 1.064 μm, which is more than 5 times that of KTP, but its damage threshold is only 40–50% of KTP. KTA is a relatively new crystal. It is all arsenate isomorph of KTP. Its nonlinearity and damage threshold are similar to KTP. However, KTA has no orthophosphate overtone absorption at 3.5 μm and its infrared absorption edge is red-shifted to about 5 μm. Therefore KTA and KNbO$_3$ can be used in OPG/OPA to extend the tuning range to 4.5 μm and 5.0 μm respectively, although no experimental work on OPG/OPA using KTA or $KNbO_3$ has yet been reported because no large crystal of KTA is available at present. It has, however, been used mostly in femtosecond OPO.

3.1.3 Materials for mid-IR output

For OPG/OPA in the mid-infrared range, $AgGaS_2$ and $AgGaSe_2$ are the best available crystals. $AgGaS_2$ is transparent from 500 μm to 12 μm. With type-I phase-matching, it can be used as an OPG with an output tunable from 1.2 to 10.0 μm when pumped at 1.064 μm.[45] It can be extended to 12 μm by type-II phase matching. $AgGaSe_2$ is transparent from 0.71 to 18 μm. The tuning range of an $AgGaSe_2$-OPO covers from 1.6 to 9.06 μm when it is pumped by a Nd:YAG at 1.34 μm and from 2.5 to 12 μm when pumped by a Ho:YLF at 2.05 μm.[39] The nonlinear optical coefficient of $AgGaS_2$ at 10.6 μm is 12–18 pm/V and it is 33–58 pm/V for $AgGaSe_2$. The laser damage thresholds of both $AgGaS_2$ and $AgGaSe_2$ are low. For $AgGaS_2$, our experiment shows that it is around 600–700 MW/cm^2 when the crystal is pumped by 35-ps pulses at 1.064 μm. With similar pump pulses, however, Elsaesser et al.[45] claimed to have observed only a surface damage on $AgGaS_2$ at an intensity as high as 25 GW/cm^2, although at 3 GW/cm^2 the front surface of the crystal would show minor damage after several thousand pulses. Most other researchers have observed a much lower damage threshold in $AgGaS_2$. The crystal often turns slightly dark after long-term irradiation with picosecond pulses at an intensity higher than 600 MW/cm^2 (800 MW/cm^2 in some cases). With such a low laser damage threshold, it is then more appropriate to use the crystal in a difference frequency generation (DFG) device than in an OPG to avoid using a high pump intensity. The damage threshold of $AgGaSe_2$ pumped at 1.0654 μm is much lower than that of $AgGaS_2$ because of the presence of two-photon absorption. To pump $AgGaSe_2$ with high intensity lasers, it is necessary to use lasers with longer wavelength such as Ho:YLF at 2.05 μm.

Both $AgGaS_2$ and $AgGaSe_2$ have high indices of refraction which lead to significant Fresnel reflection at the crystal/air interface. Anti-reflection coating can be used to reduce the Fresnel loss, however, the damage threshold of anti-reflection coating in the mid-IR is generally lower than the surface damage threshold of the crystals, further reducing the maximum pump intensity.

3.2 PUMP SOURCE CONSIDERATION

Optical parametric generation (OPG) operates on amplification from spontaneous emission. Thus a high pump intensity (typically in the order of GW/cm^2) is often needed to generate a stable output. Picosecond pump pulses are therefore preferred because of the high peak intensity reachable without damaging the crystal. The choice of the pump frequency is usually determined by the output tuning range desired and by possible laser damage of the crystal. The pump sources usually employed are fundamental and harmonic outputs of a passive/active mode-locked solid-state laser (Nt:YAG and Nd:YLF). For efficient operation, a good pump beam quality is essential.

The earliest reported successful OPG/OPA[40] system used a LiNbO$_3$ crystal in a single-pass pump configuration. The pump was a single picosecond pulse from a mode-locked Nd:glass laser with a pump energy of 1 mJ and pump intensity of 10 GW/cm^2. The Nd:glass laser was later replaced by a Nd:YAG oscillator-amplifier system with a much higher energy and repetition rate.[41] The modern commercial passive/active mode-locked Nd:YAG laser can generate 30-ps pulses at 1.064 μm with an energy more than 70 mJ/pulse and a repetition rate between 10 to 30 Hz. Mode-locked Nd:YAG or Ti:Sapphire lasers with a regenerative amplifier and their harmonic output can also be used. Such systems can produce an output of 10^1–10^2 mJ/pulse in the near IR with a pulse duration of 2-3 ps (100–200 fs) and a pulses repetition rate of 10–50 Hz.

A LiNbO$_3$ OPG/OPA system pumped at 1.064 μm provides a tunable output in the near IR range between 1.5–4.0 μm (2500–6700 cm^{-1}).[40] The tuning range could be extended to 588 nm (17,000 cm^{-1}) with the pump at 532 nm, but because of the limitation on the pump intensity imposed by much lower laser damage of LiNbO$_3$ at 532 nm, the OPG/OPA output is rather weak. KTP has the same difficulty. For OPG/OPA system tunable in the visible, BBO or LBO crystals should be used. Both the second and the third harmonic output of high-power picosecond Nd:YAG laser pulses have been employed as the pump for such systems. With proper design, second harmonic generation can have an energy conversion efficiency of 60–70%[28] and third harmonic generation can have a conversion efficiency of 50%.[43]

3.3 GEOMETRIC CONFIGURATION OF OPG/OPA

We consider here an OPG/OPA system which is capable of delivering high-energy tunable output pulses. The early device[13,20,40] used the "single pass traveling wave" configuration with the pump beam and the generated beams propagating collinearly through a long crystal once only as shown in Fig. 3.1(a). The pump intensity was a few GW/cm^2. Presumably because of the poor quality of the 1.064 μm pump beam from the mode-locked Nd:glass laser, the output efficiency was only ~1% even with a 50-mm long LiNbO$_3$ crystal. With the same pump intensity from a modern picosecond laser, the corresponding conversion efficiency would be more than 10% in such a single-pass configuration. Also, the output bandwidth of the

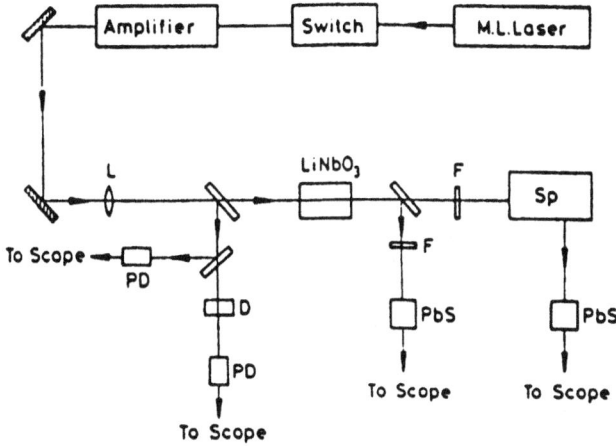

a

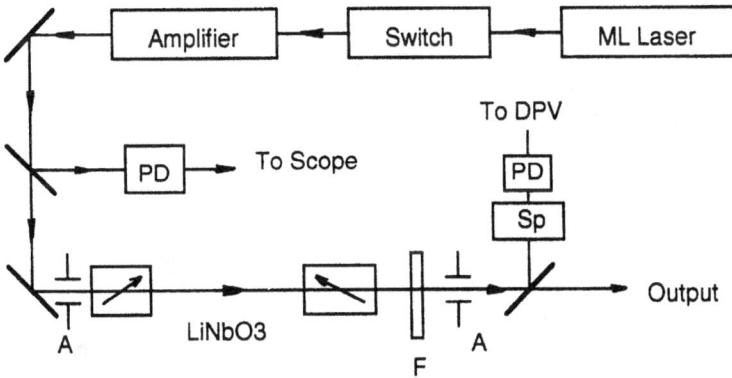

b

Fig. 3.1 Schematics of the experimental OPG/OPA systems of (a) a traveling wave optical parametric generator (OPG) using a single LiNbO$_3$ crystal [after Laubereau *et al.*, *Appl. Phys. Lett.* **25**, 87 (1974)] and (b) an improved traveling wave OPG device using two LiNbO$_3$, crystals which increased the output energy and reduced the bandwidth [after A. Seilmeier *et al.*, *Opt. Comm.* **24**, 237 (1978), *Appl. Phys.* **23**, 113 (1980)].

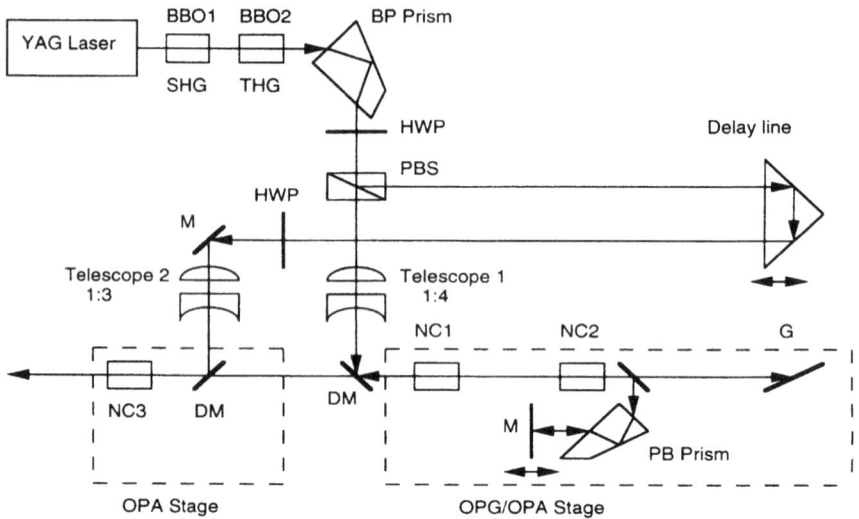

Fig. 3.2 Schematic of the experimental setup for a BBO OPG/OPA system pumped at 355 nm that has a maximum optical conversion efficiency of 30% (signal only). [After J.Y. Huang *et al.*, *Appl. Phys. Lett.* **57**, 1961 (1990)].

first OPG was rather broad ($\sim$100 cm^{-1}). Later, an improved configuration (shown in Fig. 3.1(b)) with two 50-mm long LiNbO$_3$ crystals was employed to improve the bandwidth.[31,41,42] Recently, a double-pass configuration was introduced which is capable of generating a higher output and a narrower bandwidth.[29] This configuration is the basic scheme now commonly used in OPG/OPA systems. We therefore describe such a configuration in more detail here.

Presented in Fig. 3.2 is a double-pass BBO OPG/OPA system pumped by picosecond pulses at 355 nm obtained from an active/passive mode-locked Nd:YAG laser system.[29] The pump beam is split into two parts; one part to pump a double-pass OPG/OPA stage and the other to pump an OPA stage. The pump beam for the OPG/OPA stage is telescoped down to reach a pump intensity of $\sim$3 GW/cm^2 in order to generate a sufficiently strong output from the stage. The two BBO crystals in this stage are separated by 20-cm or more and mounted in opposite orientation so that the beam walk-off effect through the crystals is minimized. For output bandwidth narrowing, the end mirror M in the OPG/OPA is replaced by a grating (for better resolution, the beam should be expanded by a telescope before directing onto the grating). The output of the OPG/OPA is then further power amplified in a third BBO crystal. For better output efficiency, the division of the pump pulse energy between the OPG/OPA and OPA stages should be judiciously chosen, depending on

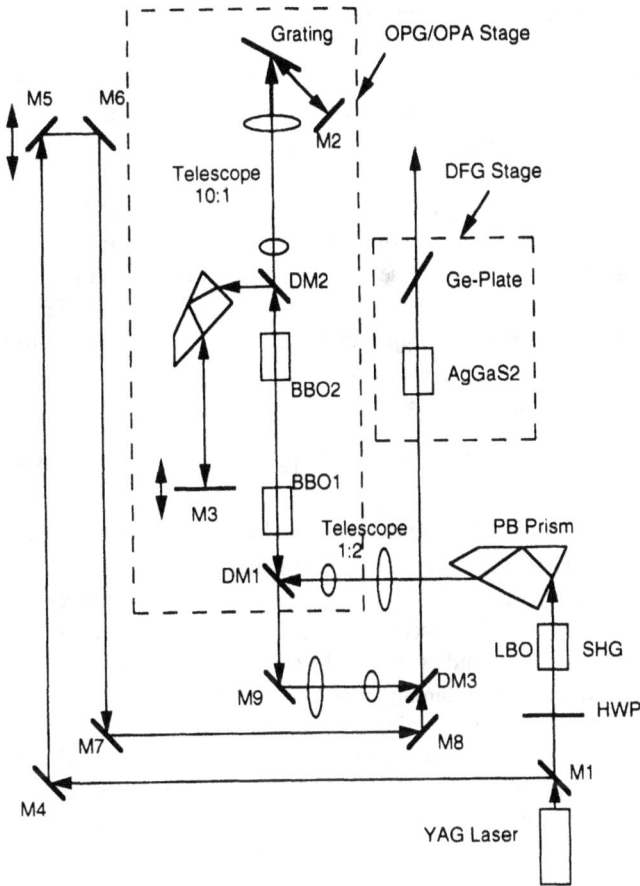

Fig. 3.3 Schematic of the experimental arrangement for a BBO-OPG/OPA system pumped at 532 nm, which is tunable from 650 nm to 2.5 μm and has an output bandwidth near the transform limit, and the wavelength extension to mid-IR (3–8 μm) via DFG in a AgGaS$_2$ crystal. [After J.Y. Zhang et al., *Nonlinear Optics 1992 Technical Digest Series* Vol. 18, PD3 and *J. Opt. Soc. Am.* **B10**, 1758 (1993)].

the total pump energy available and the crystal lengths. With a 15 ps, 3 mJ pump pulse at 355 nm, an output efficiency (signal and idler) >30% has been demonstrated in such a system with crystal lengths ranging from 6 mm to 15 mm. In cases where the available pump energy is sufficiently high and the crystal length sufficiently long, the separate OPA stage is not needed. This is the case, for example, with the second harmonic output of an active/passive mode-locked Nd:YAG laser as the pump. With a 1800-lines/mm-grating for feedback and a beam diameter of $\sim$3 mm on the grating, the output bandwidth can be less than 5–6 cm^{-1} throughout the entire tuning range. The bandwidth can be improved by an order of magnitude to about 0.5 cm^{-1} with a telescope and a double diffraction geometry shown in Fig. 3.3, although the alignment then becomes more critical.

The BBO OPG/OPA system is angle-tuned. With the pump at 532 nm, the output can be tuned from 650 nm to 2.5 μm. With the pump at 355 nm, the tuning range is from 410 nm to 2.5 μm. In the set-up, the crystals and the grating are mounted on rotation stages driven by computer-controlled stepping motors. The two crystals in the OPG stage are rotated in opposite directions in order to minimize the beam walk-off. The pump was p-polarized. The effective second-order nonlinearity of BBO in the type-I phase matching geometry pumped at 355 nm is $d_{\text{eff}}^{(2)} = 1.4$ pm/V and the figure of merit, defined as $F = \left| d_{\text{eff}}^{(2)}/n^3 \right|^2$ with n being the refractive index, is 4.9×10^{-25} m^2/V^2.

The BBO crystals in Fig. 3.2 can be replaced by LBO. The effective second-order nonlinearity of LBO is $d_{\text{eff}}^{(2)} = 1.1$ pm/V, compared to $d_{\text{eff}}^{(2)} = 1.4$ pm/V for BBO. The figure of merit of LBO is about 60% of that for BBO. For this reason, the pump intensity for onset of superfluorescence for LBO is about 2.6 GW/cm^2, as compared to 1.0 GW/cm^2 for BBO, and is generally significantly higher for the OPG/OPA operation. However, the laser damage threshold of LBO is also higher, which allows pumping of LBO at higher intensities. This partially offsets the disadvantage of the lower nonlinearity. The other advantages of LBO are its large acceptance angle ($\sim$9 mrad cm$^{1/2}$ for angle tuning and 42 mrad cm$^{1/2}$ for temperature tuning with noncritical phase matching) and low walk-off angle. They allow the use of a more highly focused pump beam and the use of a long crystal length with small cross-section.

It is also possible to extend the output tuning range of an OPG/OPA system further into the UV to $\sim$200 nm by second harmonic and sum-frequency generation, or to mid-IR around 18 μm via difference frequency generation (DFG). The experimental arrangement of the latter is shown in Fig. 3.3. Details will be discussed later.

4. EXPERIMENTAL RESULTS AND COMPARISON WITH THEORY

In this section we first present the tuning curves of OPG/OPA systems pumped at different wavelengths using various crystals. We then discuss the energy exchange between the pump beam and the generated beams inside an OPA. We later describe the bandwidth narrowing techniques used in various devices and the bandwidths obtained experimentally. We also discuss experimental results achieved on extension of the tuning range of an OPG/OPA output using frequency doubling and DFG. Finally, we summarize the performances of some practical OPG/OPA systems.

4.1 FREQUENCY TUNING

The frequencies of the signal and idler output of an OPG/OPA system are determined by the collinear phase matching condition of Eq. (9) in Section 2. They can be tuned either by crystal rotation or by temperature variation. Here, we summarize the tuning curves for various crystals that have been verified by experiments.

4.1.1 OPG/OPA pumped at 1.064 μm (or 1.053 μm)

The tuning curves of angle-tuned OPG/OPA using LiNbO$_3$,[40] KTP[46] and AgGaS$_2$[45] crystals pumped at 1.064 μm (or 1.053 μm) are presented Figs. 4.1, 4.2 and 4.3, respectively. For LiNbO$_3$, the tuning range is from 2500 cm^{-1} (4 μm) to 7500 cm^{-1} (1.5 μm), occuring at crystal orientation between 43.5° and 47° with respect to the crystal c-axis. For KTP, it is from ~4.5 to 1.5 μm for crystal orientation between 43.5° and 76°. For AgGaS$_2$, it is from 1.2 to 10 μm for crystal orientation between 35° and 55°.

In principle, other nonlinear crystals, such as BBO, LBO and KDP (KD*P) can also be used with pumping at 1.064 μm. However, their tuning ranges are more limited by absorption in the near IR. Consequently, they are seldom employed.

4.1.2 OPG/OPA pumped at 532 nm (or 526 nm)

The tuning curves of LiNbO$_3$,[31,41,42] KTP,[46] and LBO[47,51] pumped at 532 nm (or 523 nm) are given in Figs. 4.4, 4.2(b–d), and 4.7, respectively. Both angle and temperature tuning can be conveniently used for LiNbO$_3$ and LBO. In the case of LiNbO$_3$ the tuning range of the signal frequency is from 17,000 cm^{-1} (~0.59 μm)

25

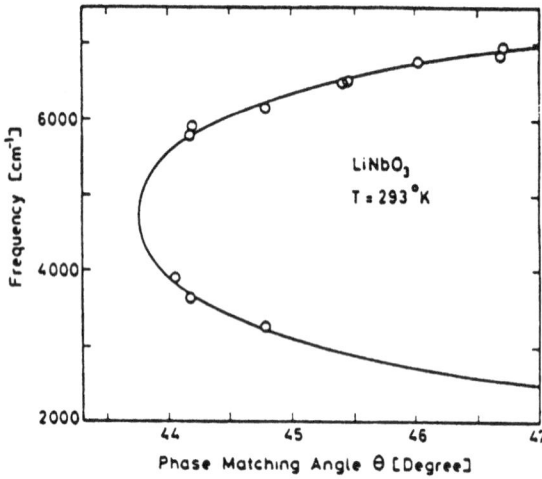

Fig. 4.1 The angle tuning curve of a picosecond type-I LiNbO$_3$-OPG/OPA system pumped at 1.064 μm. The solid curve is from theoretical calculation and the circles are experimental data. [After A. Laubereau et al., Appl. Phys. Lett., **25**, 87 (1974)].

to 11,000 cm^{-1} ($\sim$0.91 μm) when the crystal orientation changes from 50° to 90° with respect to the c-axis. (The idler can be tuned accordingly from 2,500 cm^{-1} (4.0 μm) to 7,900 cm^{-1} (1.27 μm). Temperature tuning with NCPM is possible for LiNbO$_3$ pumped at 532 nm. Varying temperature from $-14°$ to 23°C leads to an output tunable from 7,900 cm^{-1} ($\sim$1.27 μm) to 11,000 cm^{-1} ($\sim$0.91 μm). A broader tuning range should be possible by extending to higher temperatures.

For efficient wave mixing processes it is well known that noncritical phase matching is preferred because of the large acceptance angle, the relative insensitivity to beam alignment, the less stringent requirement on the beam quality and the ease of obtaining a high conversion efficiency. Thus for OPG/OPA, one would choose temperature tuning with NCPM if possible. It happens that in the case of KTP and LBO the effective nonlinearity, d$_{\text{eff}}$, is also maximum at NCPM, further improving the output. There is, however, a disadvantage inherent with temperature tuning. The thermal conductivity of nonlinear optical crystals is usually low, so that the thermal response times are long. It often takes minutes for a crystal to adjust to a new equilibrium temperature. Therefore temperature tuning of an OPG is an extremely slow process. Moreover, rapid change of the crystal temperature may induce thermal stress in the crystal. Also it is almost impossible to read accurately the output wavelength during the temperature scan of the crystal. For applications that require a reasonably fast frequency scan rate, angle-tuning of OPG/OPA is the viable scheme.

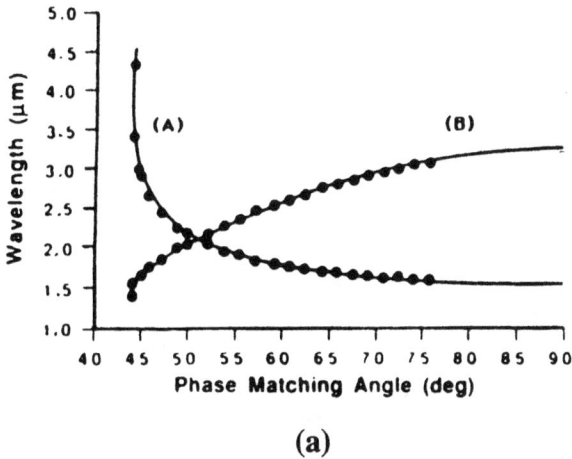

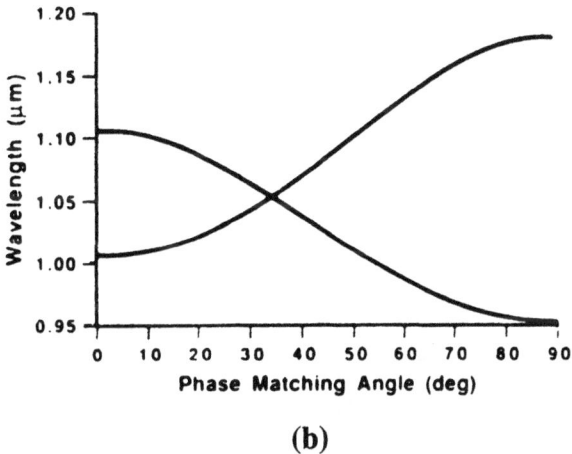

Fig. 4.2 Angle tuning curve for a type-II phase-matched KTP-OPG/OPA system pumped at 1.053 μm with beam propagation in the x–z plane ($\phi = 0$ and θ varied). Two kinds of type-II interaction can take place as indicated by A and B. For curve A the idler is polarized as an ordinary beam, while the corresponding signal is extraordinary. For curve B, the idler is the extraordinary beam, while the signal is ordinary. (b), (c) and (d) are the tuning curves, while the pump is at 526 nm and the propagation is respectively in the x–y plane ($\theta = 90°$), the y–z plane ($\phi = 90°$), and the x–z plane ($\phi = 0°$). The dots represent experimental data. [After H. Vanherzeele, *Appl. Opt.* **29**, 2246 (1990)].

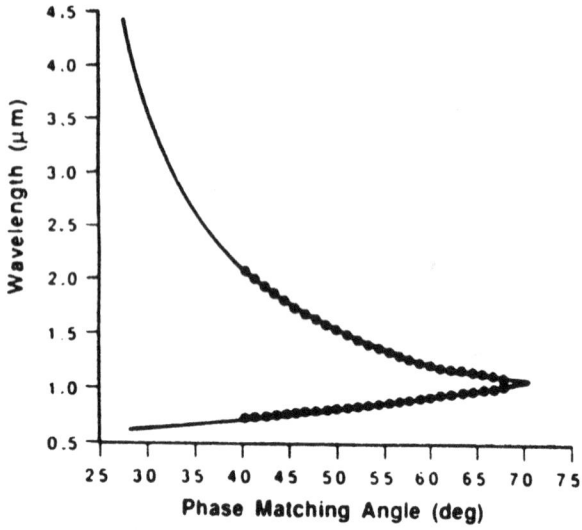

(c)

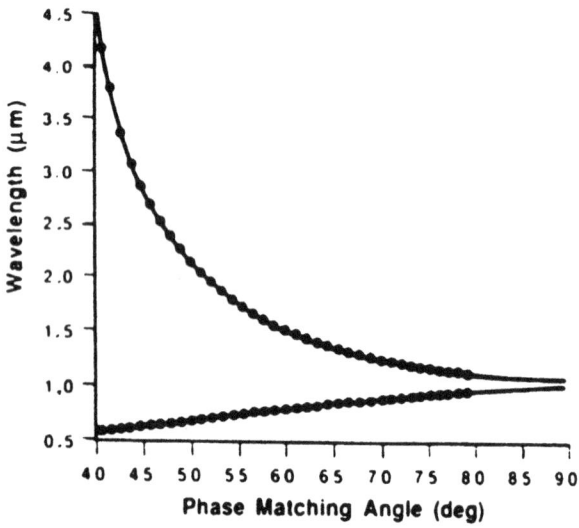

(d)

Fig. 4.2 *Continued.*

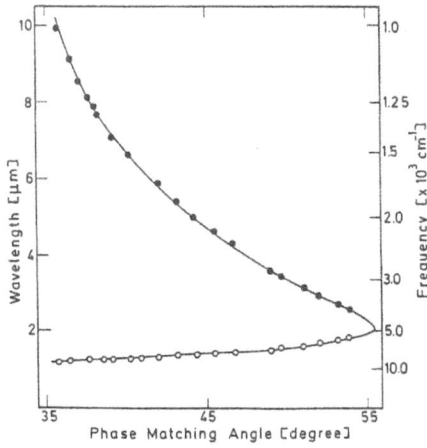

a

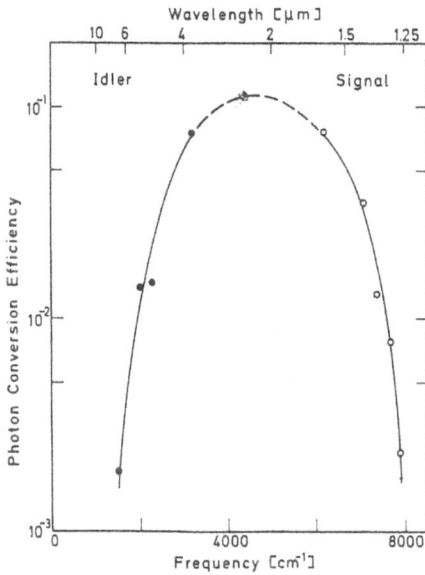

b

Fig. 4.3   (a) Angle tuning curve for a type-I AgGaS$_2$-OPG/OPA system pumped at 1.064 μm and (b) photon conversion efficiency as a function of the output frequency of the OPG/OPA. The pump energy is 10 mJ and pump intensity is 3 GW/cm^2. [After T. Elsaesser *et al.*, *Appl. Phys. Lett.*, **44**, 383 (1984)].

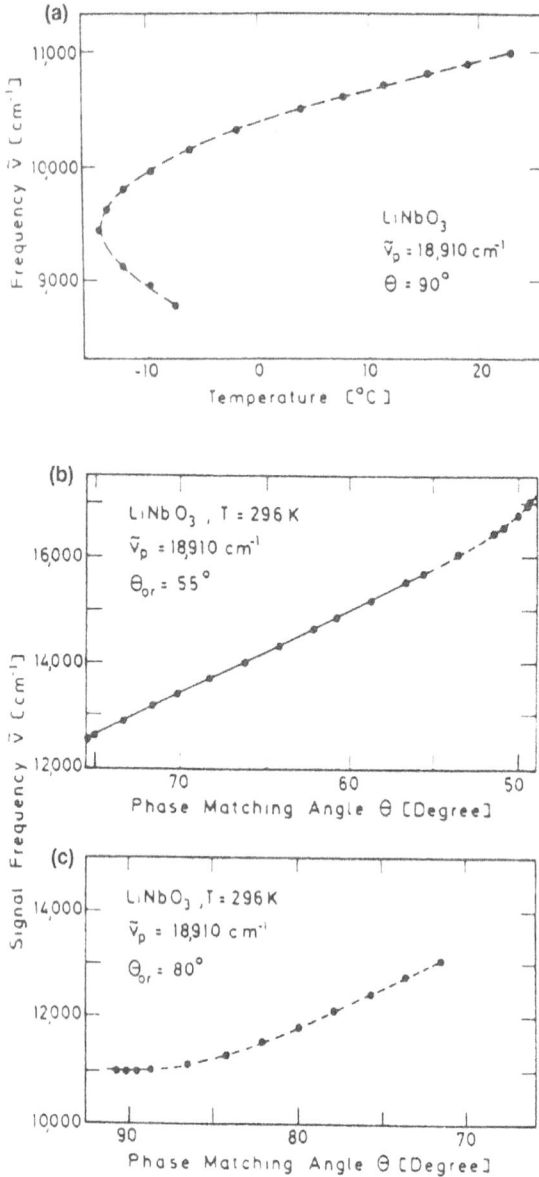

Fig. 4.4 The tuning curves for a type-I phase-matched LiNbO₃-OPG/OPA system pumped at 532 nm with (a) temperature tuning, and (b) and (c) angle tuning of signal frequency only, respectively. [After A. Seilmeier *et al.*, *Appl. Phys.*, **23**, 113 (1980)].

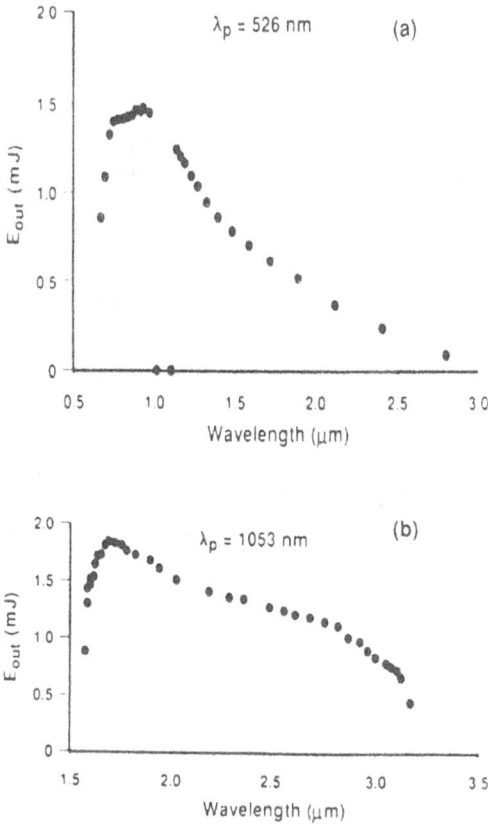

Fig. 4.5   (a) Output energy of the KTP-OPG/OPA system pumped at 526 nm as a function of wavelength. The pump energy is 12 mJ. (b) Output energy for the KTP-OPG/OPA device as a function of wavelength using phase matching scheme A as shown in Fig. 4.2(a). The pump wavelength is at 1053 nm and the pump energy is 28 mJ. [After Vanherzeele, *Appl. Opt.*, **29**, 2246 (1990)].

For KTP, the crystal is biaxial. The frequency tuning can be achieved by varying either the polar angle θ with respect to the C-axis or the azimuthal angle ϕ in the plane perpendicular to the C-axis. The tuning range is broader with θ as a variable. As shown in Fig. 4.2(b–d), varying θ, with ϕ fixed at either 0° or 90°, can yield a tuning range from 0.6 to 4.5 μm. However, due to the absorption at wavelengths longer than 3.5 μm, the useful tuning range is from 0.6 to 3.5 μm. The output energies of a KTP-OPG/OPA system pumped at 526 nm and 1053 nm as functions of wavelength are shown in Fig. 4.5(a) and (b) respectively.

For angle-tuned BBO, the tuning curve extends from 0.65 to 2.6 μm, when the crystal orientation varies between 20.7° to 22.8°. The angle tuning curve for a type-I BBO-OPG OPA system pumped at 532 nm and its output energy as a function of wavelength are shown in Fig. 4.6(a) and (b) respectively. Since the output frequency is extremely sensitive to the crystal orientation, the output bandwidth is expected to be very broad. This could be a disadvantage. However, as we discussed earlier, a bandwidth narrowing scheme can be easily incorporated in an OPG/OPA system.

For LBO, the crystal is also biaxial. The tuning range is from 0.65 to 2.7 μm, which can be obtained either by angle tuning of θ from 6.5° to 16° with ϕ = fixed, or by temperature tuning with NCPM ($\theta = 90°$, $\phi = 0°$) from 100°C to 225°C. The tuning curves of LBO exhibit a peculiar behavior as shown in Fig. 4.7, i.e. the signal frequency of the parametric output does not increase or decrease monotonously with the crystal rotation or temperature. In some regions, the output of OPG can have simultaneously two sets of signal and idler frequencies determined by the phase matching condition.[47] The appearance of this retracing behavior in LBO can be understood as follows. Let the three optical axes of LBO be X, Y and Z, and consider the case of Fig. 4.7 where all beams propagate in the X–Y plane, but at an angle from X. The pump beam is polarized perpendicular to Z and at angle ϕ from Y, and the signal and idler beams are polarized parallel to Z. The phase matching condition is written as

$$\Delta n + [n_1(\omega_p) - n_Y(\omega_S)] = 0 \qquad (14)$$

where

$$\Delta n = n_Y(\omega_p) - n_z(\omega_s)\omega_s/\omega_s - n_z(\omega_i)\omega_i\omega_p$$

$$1/n^2(\omega_p) = \sin^2\phi/n_x^2(\omega_p) + \cos^2\phi/n_y^2(\omega_p) \qquad (15)$$

Using the Seilmeier equation for LBO[47]

$$n_x^2 = 2.45316 + 0.01150/(\lambda^2 - 0.01058) - 0.01123\lambda^2$$

$$n_y^2 = 2.53969 + 0.01249/(\lambda^2 - 0.01339) - 0.02029\lambda^2$$

$$n_z^2 = 2.58515 + 0.01412/(\lambda^2 - 0.00467) - 0.0179132\lambda^2$$

$$- 4.17241 \times 10^{-4}\lambda^4 + 7.65183 \times 10^{-6}\lambda^6 \qquad (16)$$

we can calculate Δn as a function of the signal wavelength λ, as shown in Fig. 4.8. It is seen that Δn exhibits a positive minimum of about 0.008 at 0.71 μm. Then, with $[n_1(\omega_p) - n_y(\omega_p)] < -0.008$ for a certain range of θ, Eq. (14) can be satisfied by two different values of ω_S at the same ϕ. The same is true for noncritical phase

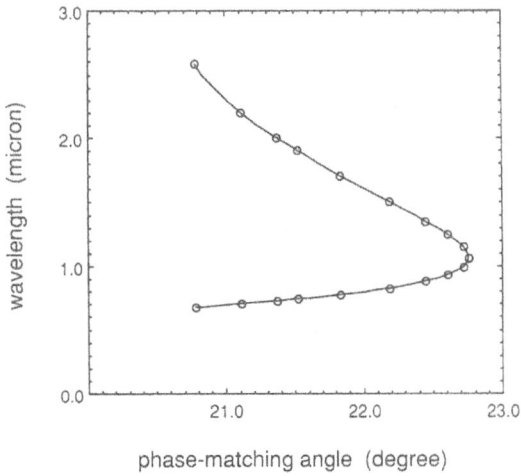

a

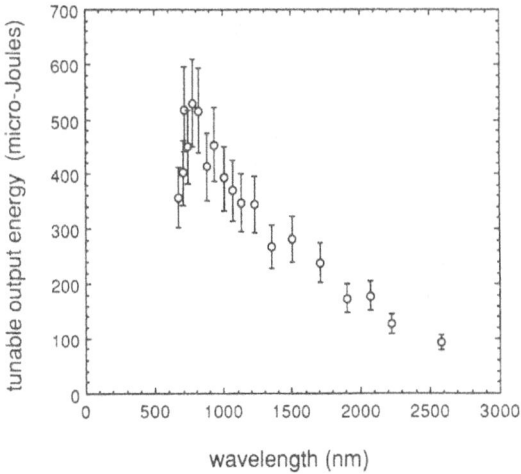

b

Fig. 4.6 (a) Angle tuning curve for a type-I BBO-OPG/OPA system pumped at 532 nm; (b) output energy of the system versus output wavelength. The pump energy in the OPA is 4.2 mJ. [After X.D. Zhu *et al.*, *Appl. Phys. Lett.*, **61**, 1490 (1992)].

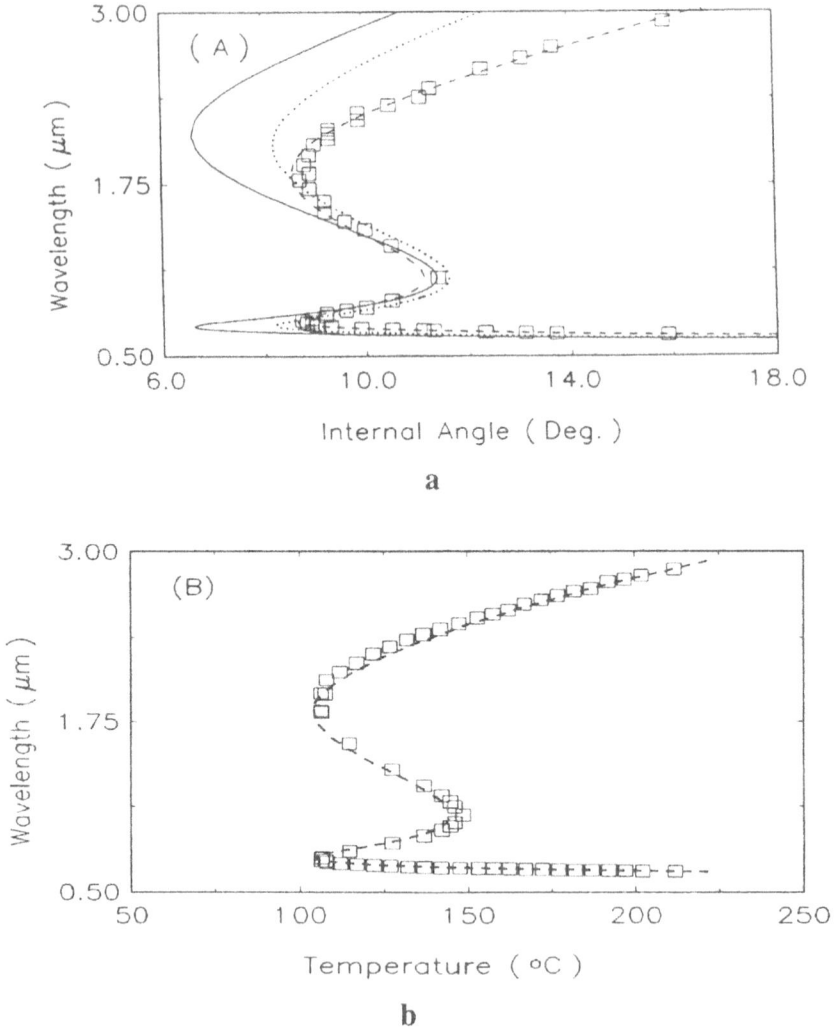

Fig. 4.7 The tuning curves of a type-I LBO-OPG/OPA system ($e \rightarrow o + o$) pumped at 532 nm via (a) angle tuning (internal angle is the angle between the surface normal and the optical X axis of the crystal) and (b) temperature tuning. The squares are the experimental data, and the solid and the dashed curves in (a) are from the phase matching calculation using modified Seilmeier equations (see reference [47]). The dotted curve is calculated from the Seilmeier equation given by K. Kato, *IEEE J.*, **QE-26**, 1173 (1990). A retracing behavior is seen in both temperature- and angle-tuning curves. [After S. Lin *et al.*, *Appl. Phys. Lett.*, **59**, 2805 (1991)].

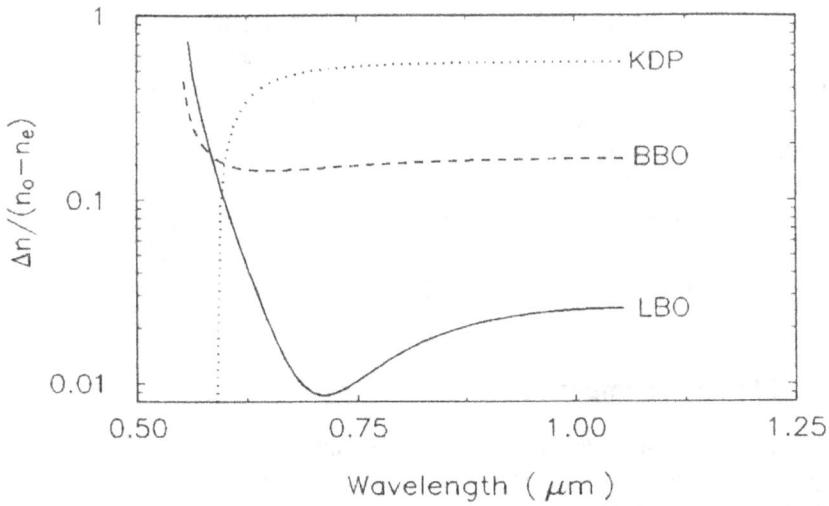

Fig. 4.8 Calculated values of Δn defined in Eq. (15) as a function of signal wavelength for LBO (solid curve), BBO (dashed curve) and KDP (dotted curve) pumped at 532 nm. [After S. Lin *et al.*, *Appl. Phys. Lett.*, **59**, 2805 (1991)].

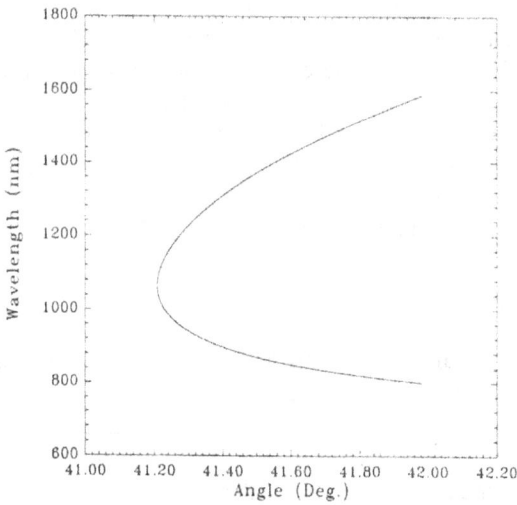

Fig. 4.9 The calculated angle-tuning curve of a type-I phase-matched KDP-OPG pumped at 532 nm.

matching in LBO when the temperature is sufficiently high. Such a retracing behavior is not found in BBO and KDP when they are pumped at 532 nm. This is because, as described in Fig. 4.8, Δn vs. ω_S for BBO has a very broad minimum and that for KDP has no minimum at this pumping frequency. In principle, the retracing behavior observed in LBO can be seen in most other nonlinear crystals if one can select the pump frequency properly.

The calculated tuning range of a type-I phase-matched KDP-OPG pumped at 532 nm is from 800 to 1600 nm (limited by the absorption in the near infrared) as shown in Fig. 4.9. Kabelka *et al.*[48] have reported the operation of a KDP OPG/OPA system with an output conversion efficiency of more than 50% at the signal wavelength of ~900 nm.

4.1.3 OPG/OPA pumped at 355 nm

The tuning curves of angle-tuned BBO-, LBO- and KDP-OPG pumped at 355 nm are presented in Figs. 4.10, 4.11 and 4.12. For BBO, the tuning range is from 0.42 to 2.28 μm when the crystal orientation is varied between 26° and 33°, limited by the cross-section of the crystal. It can probably be extended to close to 0.41 and 2.6 μm by rotating the crystal to a smaller angle. For LBO, the tuning range is from 0.405 μm to 2.85 μm when the angle ϕ varies between 18° and 42°.[30,43] The wavelength tuning becomes insensitive to the temperature of the crystal with NCPM in LBO pumped at 355 nm and can only be tuned over a narrow range.

KD*P and KDP are also good nonlinear crystals for OPG/OPA pumped at 355 nm.[39,49] The tuning range of a type-II phase-matched KDP is from 450 to 650 nm and from 800 to 1600 nm with the crystal orientation varied between 48° and 74° as shown in Fig. 4.12(a). The tuning range of a type-I KD*P-OPG/OPA pumped at 355 nm is about the same as that of KDP. Baumgartner and Byes[39] have demostrated a type-I KD*P-OPA system for a dye laser with a tuning range from 532 to 1064 nm when the orientation of the crystal was rotated from 46.4° to 49°. A KD*P-OPG/OPA system with a broader tuning range should be possible.

4.1.4 OPG/OPA pumped at 266 nm

Both BBO and LBO can also be pumped at 266 nm to generated tunable parametric output in the UV and IR. It has been demonstrated that the output of an LBO OPO pumped at 266 nm is tunable around 314 nm (signal) and 1.74 μm (idler) with angle tuning.[50] Theoretically it can be tuned from 310 nm to 2.8 μm. However, a crystal rotation of more than 90° is needed, since the output frequency is not sensitive to the orientation of LBO when it is pumped at 266 nm. Temperature tuning with NCPM in LBO has also been achieved with a tuning range from 311.5 nm to 314 nm when the temperature varies from 20°C to 70°C.[50]

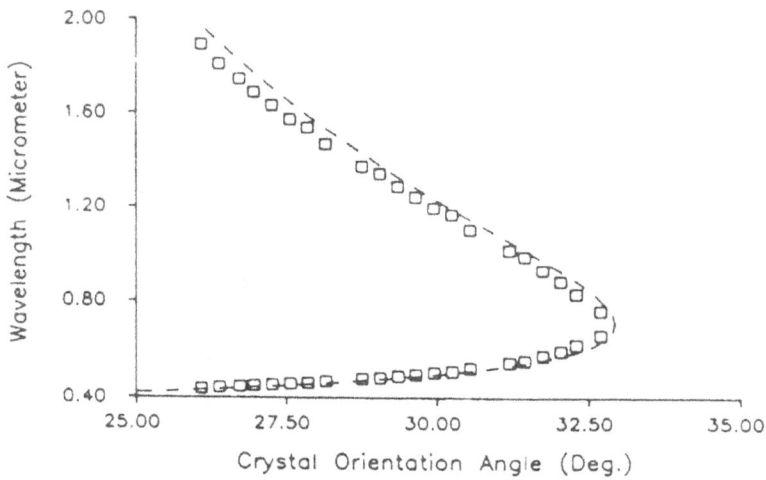

a

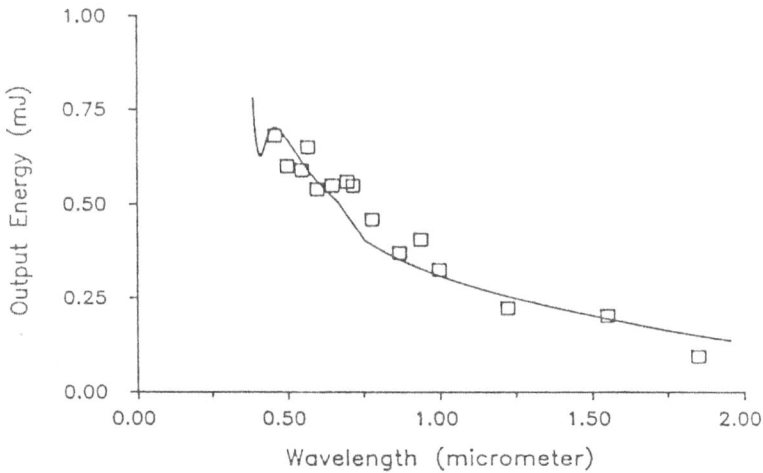

b

Fig. 4.10 (a) The angle-tuning curve for a type-I phase-matched BBO-OPG/OPA system pumped at 355 nm: the squares are the experimental data and the dashed curve is calculated; (b) the output energy of a picosecond BBO-OPG/OPA device as a function of output wavelength. The pump energies are 0.7 mJ (3 GW/cm^2) in OPG and 2.3 mJ (2.7 GW/cm^2) in OPA. [After J.Y. Huang *et al.*, *Appl. Phys. Lett.*, **57** 1961 (1990)].

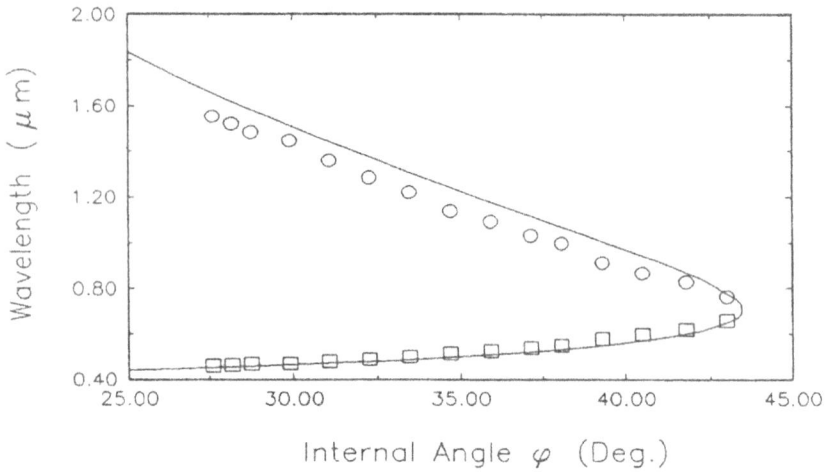

a

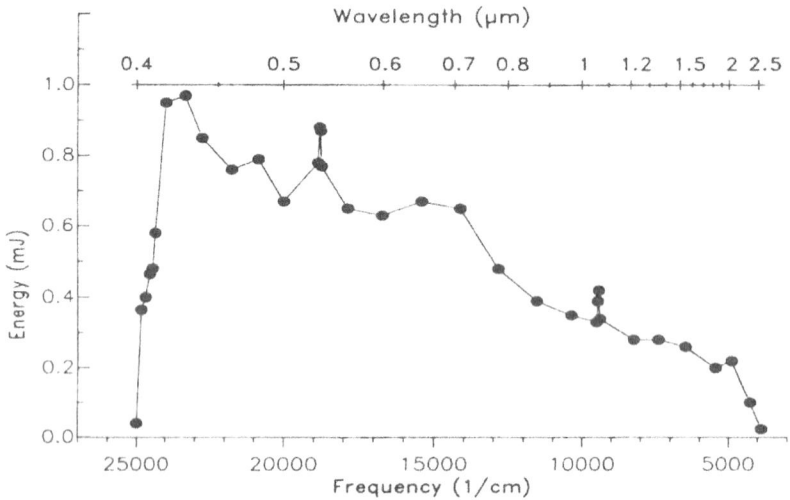

b

Fig. 4.11 (a) The tuning curve of a picosecond type-I phase-matched LBO-OPG/OPA system pumped at 355 nm, the circles (idler) and the squares (signal) are the experimental data and the curve is the theoretical result. [After J.Y. Zhang, *Appl. Phys. Lett.*, **58**, 213 (1991)]; (b) the output energy of a LBO-OPG/OPA system as a function of output wavelength. The total pump energy is 4.3 mJ [after H.-J. Krause and W. Daum, *Appl. Phys. Lett.*, **60**, 2180 (1992)].

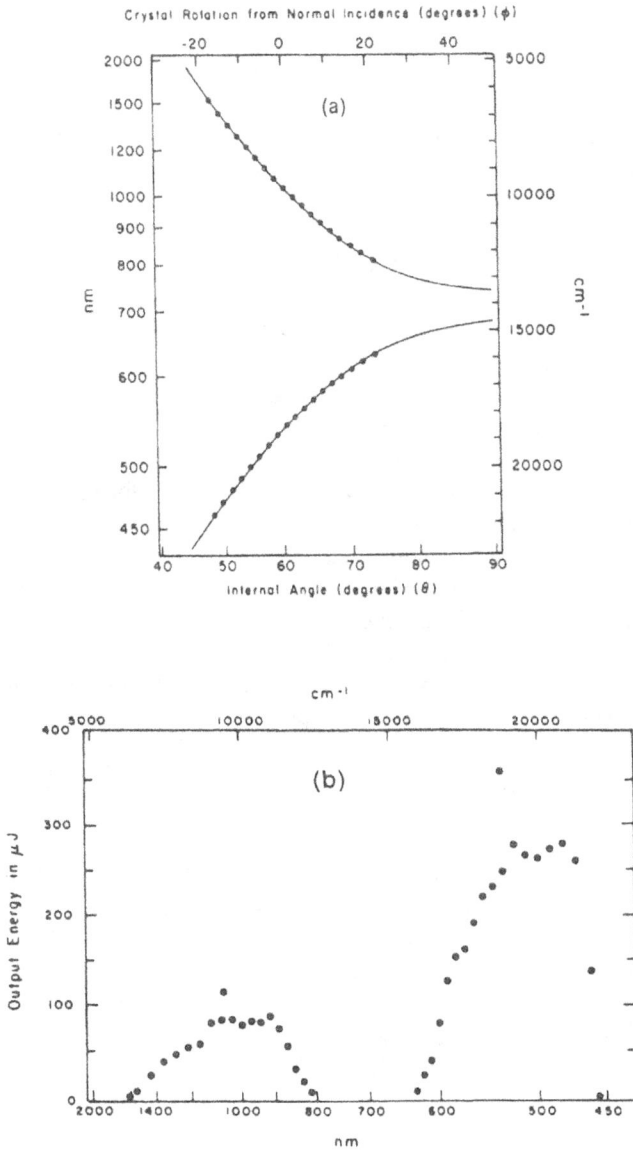

Fig. 4.12 (a) The angle-tuning curve for a picosecond type-II phase-matched KD*P-OPG/OPA system pumped at 355 nm: the dots are the experimental data and the curves are from theoretical calculation; (b) output energy of the KD*P-OPG/OPA system pumped at 355 nm. The pump energy is 1.35 mJ. [After D.W. Anthon *et al.*, *Rev. Sci. Instrum.* **58**, 2054 (1987)].

4.2 ENERGY CONVERSION IN OPG/OPA

For high-power OPG/OPA stage, the energy conversion efficiency from the pump to the output can be significant. It can reach, for example, $\sim$30% (signal only) for a BBO or LBO OPG/OPA system pumped at 355 nm. The theoretical description of how energy is converted back and forth between the pump and the signal and idler waves in an OPA has been summarized in Section 2. Here we compare experimental results with theory. In this respect, the more detailed experimental study is on BBO and LBO systems pumped at 355 nm and 532 nm.

In the case of a BBO-OPA pumped at 355 nm, the signal output versus input was measured with the 15-ps pump pulse kept at 2.3 mJ/pulse and 2.8 GW/cm^2. Figure 4.13 shows the result of the output of an OPA as a function of the injected energy at three different signal wavelengths (460 nm, 570 nm and 650 nm) obtained with a 15-mm, type-I, BBO crystal. Because the pump intensity/crystal length combination was properly chosen to yield approximately the maximum energy conversion with a significant pump depletion, the output increases only a factor of $\sim$2 when the input seed beam energy is varied over five orders of magnitude. This is in qualitative agreement with the theoretical prediction given in Fig. 2.3. The quantitative difference is presumably due to the crude approximation used in the calculation to describe the spatial and temporal profiles of the beams. The more rapid increase of the signal output occurring when the input seed beam becomes non-negligible in comparison with the output was actually observed. Operation of an OPA in the plateau region provides a more stable output as the effect of input fluctuations is greatly suppressed. A similar behavior was found in an LBO-OPA pumped at 355 nm. Measurements were earned out at the signal wavelength of 590 nm in a type-I angle-tuned LBO crystal 16 mm long. Two pump energies at 2.3 and 1.5 mJ/pulse, with peak intensities of 2.8 and 1.7 GW/cm^2, respectively, were used. The results are shown in Figure 4.14, in comparison with those obtained from the type-I, 15-mm BBO crystal. In all cases, the output increases less than a factor of 2 when the input varies from 10^{-4} to 10 μJ. We notice in Fig. 4.14 that, although d_{eff} for BBO is 1.26 times higher than that for LBO, the outputs are about the same in the two cases at the pump intensity of 1.7 GW/cm^2. In fact, when the pump depletion is significant, one can be stronger or weaker than the other depending on where the operating points are in the plot of energy conversion versus pump intensity and crystal length (see Fig. 2.2, for example). A crystal with smaller nonlinearity could generate a higher output if the crystal length is chosen properly. By the same token, if the nonlinearity is fixed and the pump intensity is properly chosen, a shorter crystal could yield a larger output. The latter possibility is seen in the theoretical plot of Fig. 4.15. Experimentally, this has been qualitatively verified by measuring the OPA outputs of three BBO crystals of different lengths, 6, 12, and 15 mm. All were pumped by 15 ps, 355-nm-pulses at pump intensity 2.8 GW/cm^2. The measured signal outputs at 590 nm were 200 μJ, 460 μJ, and

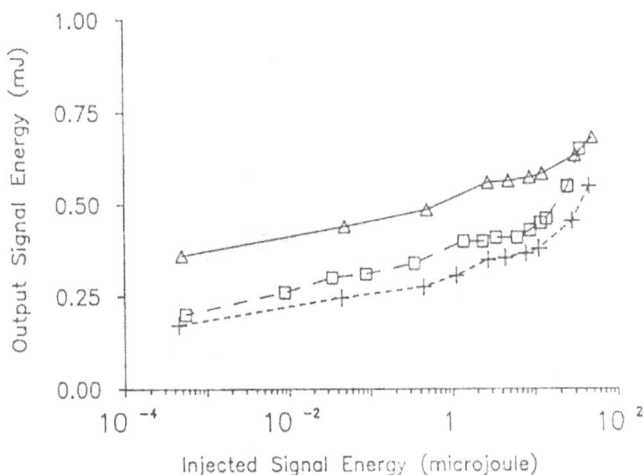

Fig. 4.13 Output energy vs. injected energy for a picosecond BBO OPA system. The 15-ps pump pulse has an energy of 2.3 mJ at 355 nm, focused to a 0.27 cm beam spot. Triangles, squares and crosses refer to output signal wavelengths at 460, 570 and 650 nm, respectively. [After J.Y. Huang *et al.*, *Appl. Phys. Lett.*, **57**, 1961 (1990)].

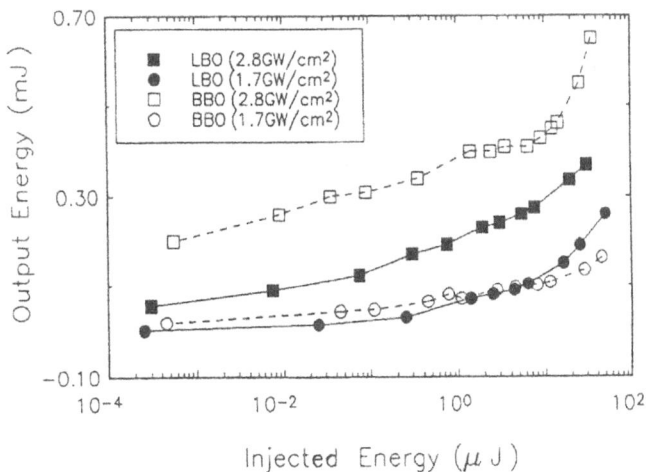

Fig. 4.14 Output energy vs. injected energy of picosecond OPA at 590 nm from a 16-mm LBO crystal (solid symbols) and a 15-mm BBO crystal (open symbols) pumped at 355 nm. Two different pump energies, 2.3 and 1.5 mJ, corresponding to 2.8 and 1.7 GW/cm^2, were used. [After J.Y. Zhang *et al.*, *Appl. Phys. Lett.*, **58**, 213 (1991)].

a

b

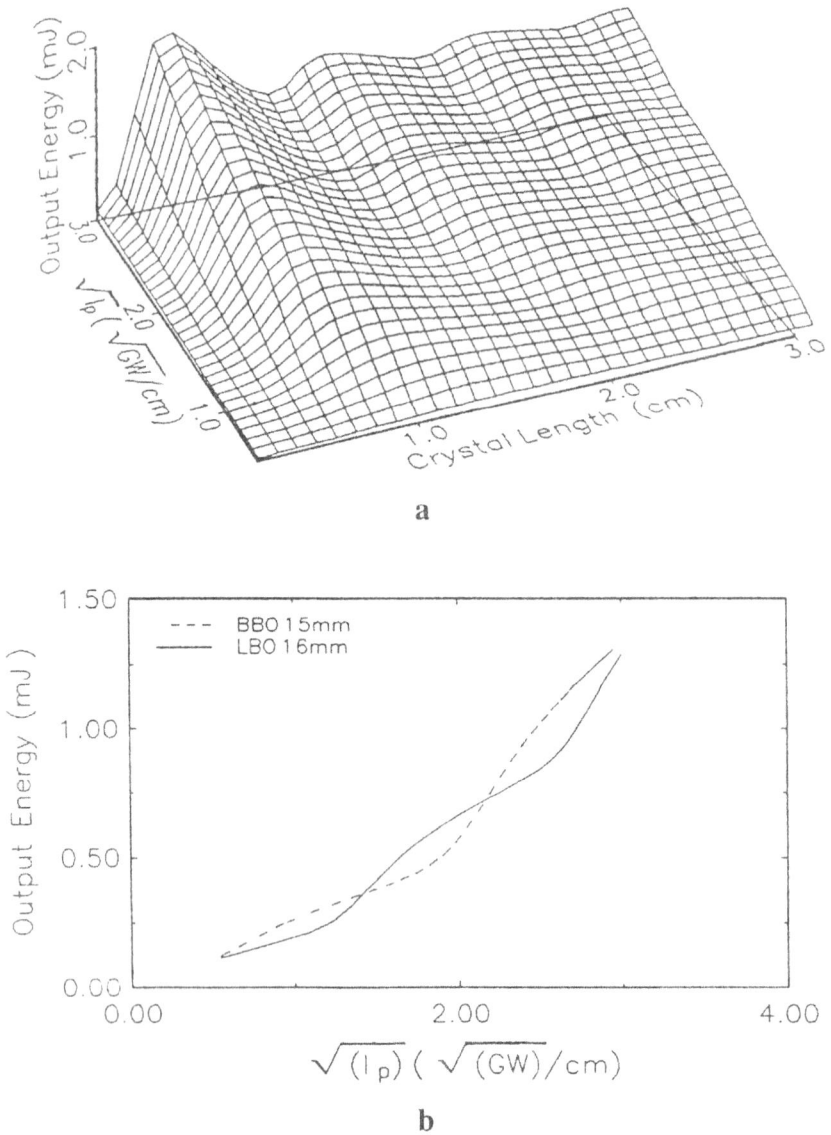

Fig. 4.15 (a) Calculated output signal energy at 590 nm as a function of crystal length L and pump intensity I_p for a type-I LBO-OPA pumped at 355 nm. (b) Output signal energy at 590 nm from a 16 mm LBO (solid curve) crystal. The injected energy and wavelength were 30 μJ and 590 nm, respectively. [After J.Y. Zhang *et al.*, *Appl. Phys. Lett.* **58**, 213 (1991)].

400 μJ, respectively. The 12 mm crystal had the highest output. With a higher pump intensity, a shorter crystal length could be better. Similar behavior was also observed in LBO. The theoretical calculation on LBO produces a plot of signal output versus pump intensity and crystal length very similar to that of BBO, as shown in Fig. 4.15(a) for comparison with Fig. 2.3. Specifically, for a given crystal length, the signal output increases with increase of I_p, but oscillates weakly. This is seen explicitly in Fig. 4.15(b) for a 16 mm LBO crystal and for a 15 mm BBO crystal. Because of the oscillations in the curves the output from LBO can be larger or smaller than that from BBO with the same pump intensity, as mentioned earlier.

We now briefly summarize the reported energy conversion in various OPG/OPA systems. We consider only systems pumped by picosecond pulses (10–30 ps) derived from passive/active mode-locked Nd:YAG or Nd:YLF lasers. With the pump at 1.064 or 1.053 μm, Laubereau and co-workers[40] found a maximum pump-to-signal energy conversion efficiency of 1% in their LiNbO$_3$ OPG/OPA system with a pump energy of 1 mJ. It could reach ~5% with better pump beam quality. For the KTP system of Vanherzeele et al.,[46] the signal output versus wavelength is given in Fig. 4.5(b). The maximum energy conversion efficiency is 6.5% at 1.7 μm. For the AgGaS$_2$ system of Elsaesser et al.,[45] the photon conversion efficiency as a function of signal wavelength is presented in Fig. 4.3(b). The maximum is ~10% (corresponding to an energy conversion efficiency of 4.8%) at 2.2 μm. Consider next OPG/OPA systems pumped at 532 nm or 526 nm. The pump-to-signal conversion efficiency of the LiNbO$_3$ system is generally low[31] because the pump intensity is limited by the optical damage. For the KTP system of Vanherzeele et al.,[46] the measured output is shown in Fig. 4.5(a). The maximum energy conversion efficiency is 12.5% at 0.9–1.0 μm. For the BBO system of Zhu et al.,[44] the output versus wavelength curve is presented in Fig. 4.6(b) with a maximum efficiency of 13% at ~750 nm. The reported maximum conversion efficiency of an LBO system is about the same.[47] For KDP, Kabelka et al.[48] reported a conversion efficiency larger than 50% with a pump intensity of 15–20 GW/cm^2. However, the crystal would be easily damaged when operating at such a high intensity. Finally, consider the pump at 355 nm. The BBO system of Huang et al.,[29] has the output as a function of wavelength as shown in Fig. 4.10(b). The maximum appears at ~460 nm with a conversion efficiency of 30% (signal only). The same conversion efficiency has also been found in the LBO systems of Krause et al. (see Fig. 4.11(b)).[43,55]

4.3 BANDWIDTH NARROWING TECHNIQUES

In many spectral applications of OPG/OPA, the bandwidth of the device is of major importance. An OPG/OPA system with a transform-limited bandwidth is highly desirable. The mechanisms of bandwidth broadening in OPG/OPA were discussed in Section 2. Here we describe briefly the experimental results and some technical details in our effort to reduce the bandwidth of an OPG/OPA system to near the transform limit.

4.3.1 Reduction of bandwidth due to collinear phase mismatch

As discussed in Section 2, bandwidth due to collinear phase mismatch is proportional to 1/L when 2gL is a constant (L is the crystal length). Therefore, it can be significantly reduced by using a long crystal. For a 4 cm LiNbO$_3$ crystal pumped at 532 nm with 2gL = 30, the calculated bandwidth ranges from 3 cm^{-1} at 15,900 cm^{-1} to 12 cm^{-1} at 11,000 cm^{-1}, and to 36 cm^{-1} at 3500 cm^{-1}. For a 1 cm BBO pumped at 355 nm with 2gL = 30, the signal output at 550 nm is estimated to have a bandwidth of 120 cm^{-1}. In a modern OPG/OPA system, two or more crystals are often used and the beams would pass the crystals twice to effectively increase the crystal length as depicted in Figs. 3.2 and 3.3. The bandwidth also increases with the pump intensity (power broadening). This will be discussed later.

4.3.2 Reduction of bandwidth due to non-collinear phase matching

As discussed in Section 2, bandwidth due to non-collinear phase matching is mainly limited by the aperture angle α in the OPG stage. Such a broadening can be significantly reduced by using two crystals in tandem, but at a significant distance apart. In a two-LiNbO$_3$-OPG/OPA device,[31,41,42] Seilmeier and Kaiser used two 5 cm long crystals oppositely oriented to compensate the walk-off and separated by approximately 30 cm. This configuration has a strong aperturing effect due to the long crystal lengths and the large separation between two crystals. The effective α is greatly reduced and hence the output bandwidth. They observed a bandwidth of 4–10 cm^{-1} for the system in comparison with 100 cm^{-1} when only one crystal was used.

4.3.3 Other bandwidth broadening mechanisms

The two bandwidth broadening mechanisms discussed in Section 2.5.1 and 2.5.2 as well as Sections 4.3.1 and 4.3.2 often dominate in a picosecond OPG/OPA system. With proper measures, the bandwidth due to these mechanisms can be reduced to 10 cm^{-1} or less without any dispersive element in the OPG stage. The contributions of other broadening mechanisms, such as divergence of the pump beam and bandwidth of the pump beam are less important in comparison with them. This is because a modern commercial mode-locked picosecond solid state laser can deliver laser pulses with a nearly transform-limited bandwidth and a nearly diffraction-limited divergence. For a high-energy OPG/OPA system requiring a nearly transform-limited output bandwidth, power broadening of the bandwidth could be significant and needs to be suppressed. This problem will be addressed below.

4.3.4 A practical system with narrow-band output

As discussed above, the bandwidth of OPG can be broad. In order to obtain a high-power tunable output with high conversion efficiency and narrow bandwidth, one must use an OPA with a narrow-band injected signal seed beam. The latter can be obtained from the output of an OPG after it is bandwidth-narrowed by a frequency-dispersive element. In this case, the bandwidth of the OPA output is mainly determined by that of the injected seed beam.[32] We have studied this possibility and the limitation of bandwidth reduction using a BBO OPG/OPA system pumped by a 532 nm beam. This system has a very large inherent output bandwidth because its signal frequency is extremely sensitive to the crystal orientation (rotating the crystal by 1.5° would result in a frequency shift of tens of thousands of wave numbers).

The configuration or OPG/OPA system is shown in Fig. 3.3. The first stage consisted of two 5 mm × 5 mm × 10 mm BBO crystals, cut for type-I phase matching and separated by more than 20 cm, the crystals could be tilted in opposite directions either independently or simultaneously. With a total pump energy of 5 mJ or higher at 532 nm and the beam diameter telescoped down to about 2.5 mm to yield a pumping intensity of 3 GW/cm^2, the single-pass output through the two crystals was about 0.5 mJ, and had a bandwidth of tens of nanometers. It was then separated from the pump beam by a dichroic mirror and, after being expanded by a 10:1 telescope, was directed onto a 1,800-grooves/mm-grating (Milton Roy-5138) with a nearly constant diffraction efficiency of about 90% from 550 nm to 1.0 μm with the beam polarization perpendicular to the grating grooves. The first-order diffracted signal beam was reflected back onto the grating to be diffracted once more. The frequency-selected beam was then fed back into the BBO crystals for a second-pass amplification. The output was dominated by the amplification of the seed beam if the seed pulse leads the pump pulse by about 1 ps, and so was the output bandwidth. With such a system, the output bandwidth observed was about 0.12–0.15 nm (about 2 cm^{-1} around 800 nm) at a pump intensity of 2.2 GW/cm^{-1}. If a single diffraction from the grating was used, the bandwidth increased to about 0.25–0.30 nm at the same pump intensity. Further improvement of the bandwidth by increasing the beam size on the grating has been considered. However, for short pump pulse duration ($\sim$20–25 ps) a larger beam size on a plane grating would result in a reflected signal pulse width longer than the pump pulse and only the part overlapping temporally with the pump pulse could be amplified.

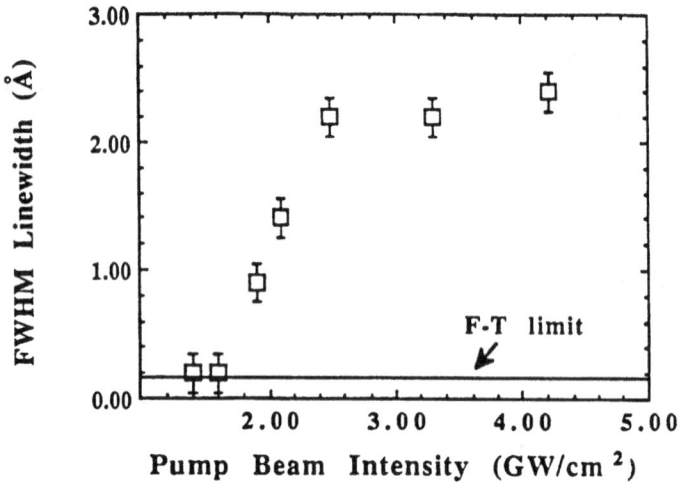

Fig. 4.16 The output bandwidth at 800 nm of a BBO-OPG/OPA system pumped at 532 nm as a function of pump intensity showing the significance of power broadening in a narrow band OPG/OPA device. [After J.Y. Zhang *et al.*, *Nonlinear Optics 1992 Technical Digest Series* Vol. 18, PD3 and *J. Opt. Soc. Am.*, **B10**, 1758 (1993)].

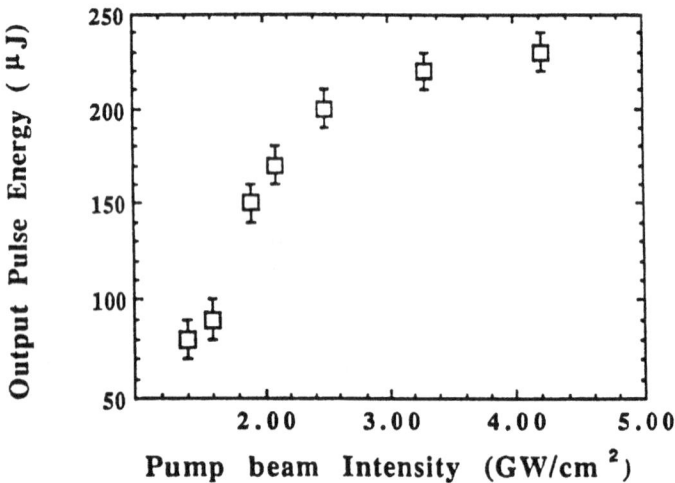

Fig. 4.17 The output energy at 800 nm of a narrow band BBO-OPG/OPA system pumped at 532 nm as a function of pump intensity. [After J.Y. Zhang *et al.*, *Nonlinear Optics 1992 Technical Digest Series* Vol. 18, PD3 and *J. Opt. Soc. Am.*, **B10**, 1758 (1993)].

As mentioned earlier, the pumping intensity can affect the bandwidth significantly. This was confirmed experimentally.[32] Figure 4.16 shows the observed output bandwidth of the above-mentioned BBO OPG/OPA system as a function of pump intensity. At a pump intensity below 1.5 GW/cm^2 the bandwidth is ~0.02 nm, which is the transform-limited bandwidth, but the output stability is poor. With increasing pump intensity, it increases rapidly to a saturated value of ~0.23 nm, corresponding to 36 cm^{-1} around 800 nm. The broadening presumably sets in when saturation of amplification of the injected seed beam becomes significant. Operating an OPG/OPA system at even higher pump intensites would make the OPA stage also an OPG and generate a broad-band parametric radiation from noise amplification, appearing as a background in the OPA output. The output energy of the OPG/OPA system as a function of pump intensity is shown in Fig. 4.17. It is seen that at pump intensity below 2 GW/cm^2, it is less than 150 μJ and the stability was found to be poor. At pump intensity above 3 GW/cm^2, the output increases to about 250 μJ. It appears saturated and hence stable.

In aligning the system, it is important to center the frequency of narrow-band seed beam at the gain maximum of the OPA. Otherwise the signal output from the OPA could have a spectrum with two peaks, one at the input frequency and the other around the frequency with the maximum parametric gain, as shown in Fig. 4.18. Reducing the pump intensity in OPA and increasing the pump intensity in OPG will emphasize the peak at the seed beam frequency and reduce the other. The output also becomes more stable and both the output bandwidth and the output energy can be optimized.

Other factors that can affect the output bandwidth are the divergence of the pump and signal beams in OPA. They become important if the frequency tuning of OPG/OPA is very sensitive to the crystal orientation as in the case of BBO pumped at 532 nm. In general, this makes bandwidth narrowing more difficult. Careful adjustment of the telescopes for the pump and seed beams to minimize the effect of beam divergence is important.

Finally, the bandwidth also depends on the signal frequency. Near the degeneracy point of the tuning curve, the bandwidth is appreciably broader as shown in Fig. 4.19. One can avoid the degeneracy point by using a different pump frequency.

4.4 FREQUENCY EXTENSION TO MID-IR BY DFG AND TO UV BY SHG

The tunable output in the visible range from an OPG/OPA system can be extended to the UV by second harmonic generation (SHG) in nonlinear crystals or to mid-IR by difference-frequency generation (DFG).

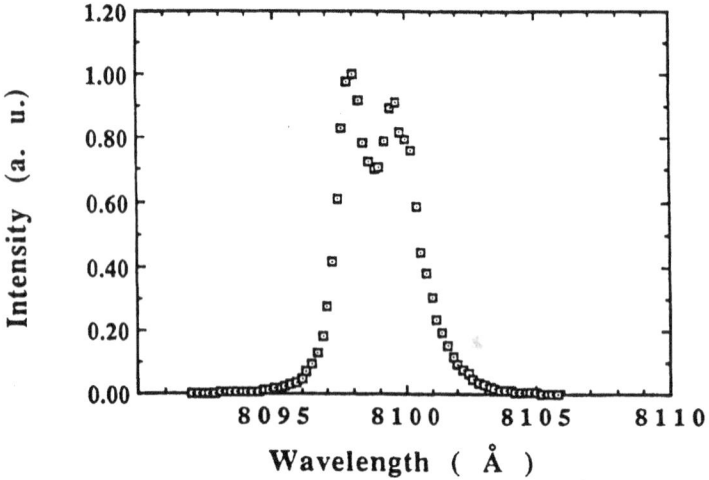

Fig. 4.18 A double-peaked output spectrum of a BBO-OPG/OPA system due to slight misalignment between the grating and BBO crystal orientation. [After J.Y. Zhang *et al.*, *Nonlinear Optics 1992 Technical Digest Series* Vol. 18, PD3 and *J. Opt. Soc. Am.*, **B10**, 1758 (1993)].

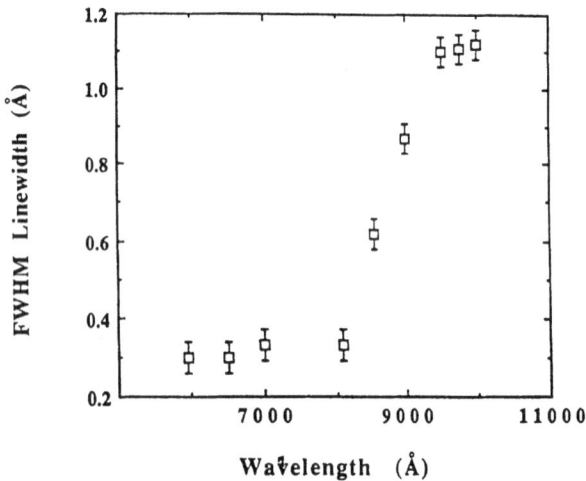

Fig. 4.19 The output bandwidth of a BBO-OPG/OPA system pumped at 532 nm as a function of signal wavelength. The pump pulse has an energy of 6 mJ and a pulse duration of 25-ps. [After J.Y. Zhang *et al.*, *Nonlinear Optics 1992 Technical Digest Series* Vol. 18, PD3 and *J. Opt. Soc. Am.*, **B10**, 1758 (1993)].

Extension of the output frequency of OPG/OPA to the UV has been demonstrated earlier[31] by frequency doubling the output of a LiNbO$_3$-OPG/OPA system up to 32,000 cm^{-1} in one LiIO$_3$(1.4 mm) long and cut at 52°) and two KDP (8 mm long and cut at 41.5° and 52°, respectively) crystals. All crystals were prepared for type-I phase matching. The energy conversion efficiency of frequency doubling was about 1%. The bandwidth of the OPG/OPA output was preserved after frequency doubling. The low conversion efficiency was presumably due to the low output power of their OPG/OPA system and to the low laser damage threshold of the crystals used for frequency doubling. With the high output power of BBO or KTP OPG/OPA systems pumped at 532 nm, one should be able to reach a 10% or higher energy conversion efficiency at wavelengths around 350–450 nm by frequency doubling in a 5 mm BBO.[65] A shorter wavelength down to 300 nm is possible. A higher conversion efficiency is expected if the BBO frequency-doubling crystal is replaced by a LBO crystal and the output beam from the OPG/OPA is more tightly focused. The tuning range can be extended to around 210 nm if the output of an LBO OPG/OPA pumped at 355 nm is frequency doubled.

To extend the tuning range to the mid-infrared, the difference frequency generation (DFG) technique can be used. The infrared output can have tuning ranges down to 3.5–14.0 μm (in LiNbO$_3$ or KTP) and to 4.5–5.0 μm (in LiIO$_3$ or KNbO^3). For mid-infrared generation, the most suitable nonlinear crystals for DFG are AgGaS$_2$ (transparent from 0.49 to 12 μm), GaSe (transparent from 0.64 to 18 μm) and AgGaSe$_2$ (transparent from 0.71 to 18 μm). It has been demonstrated that mid-infrared radiation can be directly generated by OPO pumped by a Q-switched YAG laser or by OPG/OPA pumped by a mode-locked picosecond YAG laser at 1064 nm. The tuning range is from 1.4 to 4.0 μm for OPO[52] and from 1.2 to 10 μm for OPG[53] respectively. The tunable mid-IR radiation can also be achieved by difference frequency generation (DFG).[32,54,55] Generation of narrow-band, high-power, picosecond tunable radiation in the mid-IR from 3.0 to 8.0 μm has been achieved by DFG between the laser at 1064 nm and the tunable idler output (1.65–1.22 μm) of a BBO OPG/OPA system in AgGaS$_2$. The experimental arrangement is shown in Fig. 3.3. The AgGaS$_2$ crystal used for this purpose is cut at 42° for type-I phase matching or 47° for type-II phase matching. The corresponding phase matching curves for DFG, by mixing a beam at 1.064 μm with a tunable beam in the near IR to generate an output tunable from ~12 to 4.5 μm or shorter, are presented in Fig. 4.20. They suggest that the tunable output can be extended 12 μm, as has been demonstrated recently by Krause and Daum.[55] The frequency tuning of DFG output is usually accomplished by tuning the frequency of the OPG/OPA output and rotating the AgGaS$_2$ crystal in synchronous steps via computer-controlled rotation stages to maintain phase matching of DFG. The temporal and spatial overlaps between the beams are critical and can be optimized using sum-frequency generation as a calibration. In a recent experiment, the DFG output from a 10-mm long AgGaS$_2$

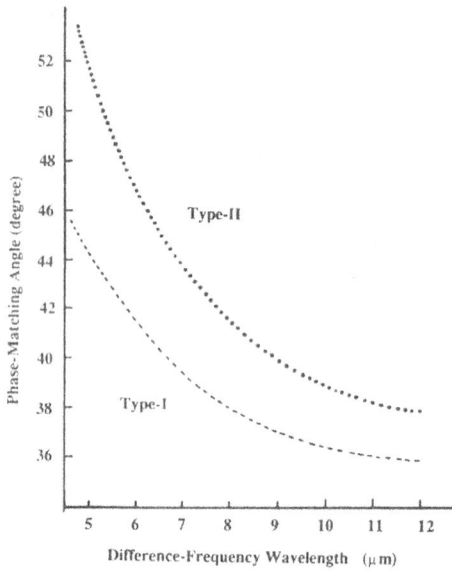

Fig. 4.20 The tuning curves of type-I and type-II phase matching for DFG in AgGaS$_2$ crystal plotted by mixing a frequency-fixed laser beam at 1.064 μm with a tunable beam in the near IR (1.17–1.39 μm) generated by an OPG/OPA system.

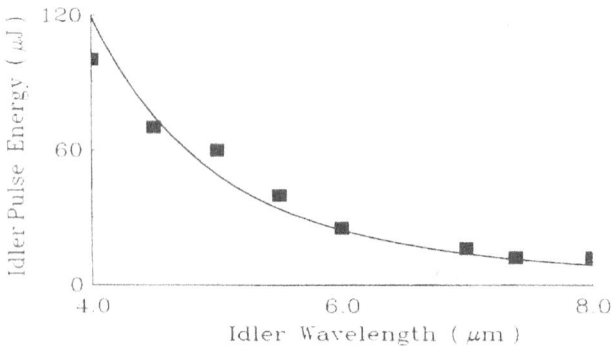

Fig. 4.21 Output energy versus output wavelength of DFG from a 10-mm type-I AgGaS$_2$ crystal. DFG results from mixing of the output of a ps Nd:YAG at 1.064 μm (3 mJ and 300 MW/cm^2) with the idler output ($\sim$10 μJ) from a 532 nm pumped BBO-OPG/OPA. The squares are the experimental data, and the solid curve is theoretical. The Fresnel loss of the crystal surface has been taken into account in the calculation. [After J.Y Zhang *et al.*, *Nonlinear Optics 1992 Technical Digest Series* Vol. 18, PD3 and *J. Opt. Soc. Am.*, **B10**, 1758 (1993)].

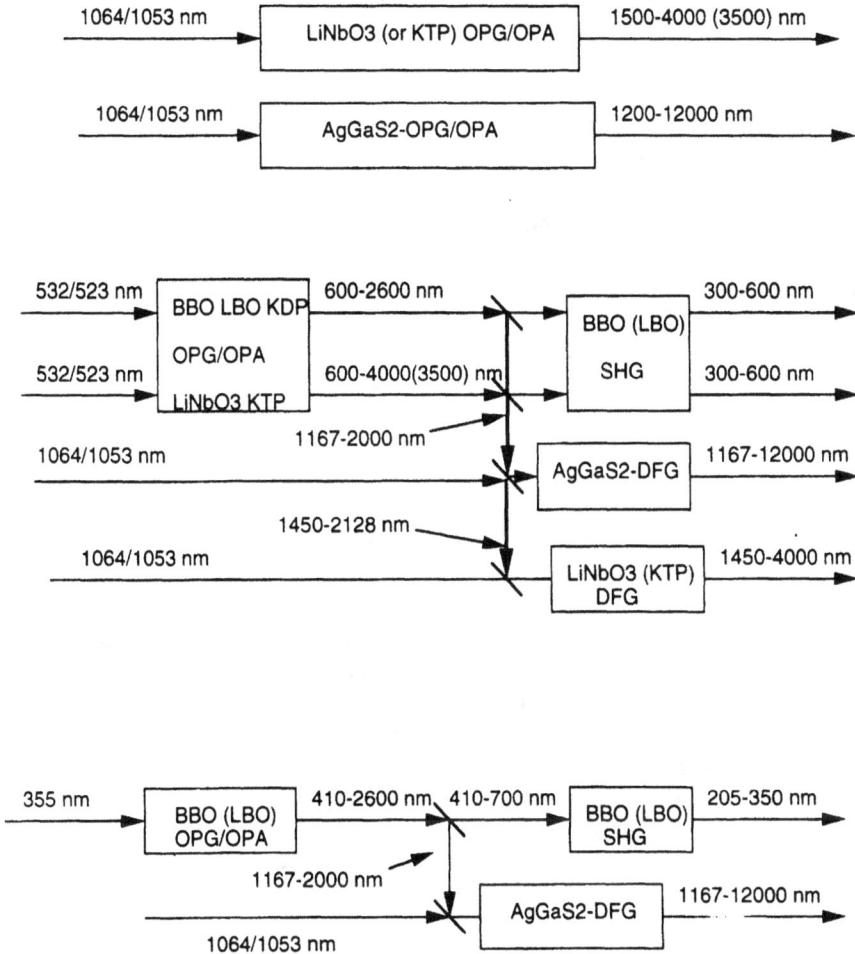

Fig. 4.22 Possible schemes for generating tunable picosecond pulses from UV to mid-IR.

crystal pumped by 3 mJ in a 6 mm spot (300 MW/cm^2 at 1064 nm and 10–20 μJ of tunable IR from the idler of an OPG/OPA system was about 120 μJ at 4 μm and decreased to 10 μJ at 8 μm, as shown in Fig. 4.21 together with the theoretical calculation.[32] The output energy at wavelengths shorter than 4 μm was not measured as it was limited by the cross-section of the crystal used. The agreement between theory and experiment is quite satisfactory. Because DFG is synonymous to OPA, its output bandwidth is expected to experience the same broadening mechanisms operative in OPA. The pump intensity in $AgGaS_2$ is usually limited by the damage threshold. Thus, for higher DFG output a longer crystal is preferred. Alternatively, two crystals in tandem can be used, as described in Ref. 55. For significant improvement of conversion efficiency, however, it is advisable to split the pump beam properly into two parts, one used to pump the first crystal and the other to pump the second crystal. Correction of phase mismatches between the waves before they enter the second crystal may be necessary for efficient energy conversion.

GaSe could also be an effective nonlinear crystal for OPO, OPA, and DFG output in the mid-IR region. It has a transparent range from 0.64 to 18 μm, a nonlinearity of 80 pm/V (four times that $AgGaS_2$), a figure of merit of 300 (ten times that of $AgGaS_2$), and a laser damage threshold similar to that of $AgGas^2$. It has been successfully used for DFG of picosecond pulses to generate mid-IR radiation tunable from 4 to 18 μm.[56] However, it is not yet commercially available and, as a layered crystal, its surface cannot be easily cut and polished along arbitrary directions. The optical surface usually can be obtained only by cleavage of the (001) plane.

Silver gallium selenide ($AgGaSe_2$) crystals have also been used to generate tunable mid-IR radiation with a tuning range from 1.6 to 6.9 μm in an OPO pumped at 1.34 μm and from 2.65 to 9.02 μm when pumped at 2.05 μm.[39] The tuning range could be extended to 18 μm in DFG. To avoid significant two-photon absorption that could lead to optical damage, the pump laser wavelength should be longer than 1.4 μm. This precludes the use of the sophisticated mode-locked Nd:YAG at 1.064 μm and Nd:YLF at 1.053 μm as the pump sources. So far, little work has been done on OPG/OPA or DFG using $AgGaSe_2$. With the improved availability of mode-locked, high-energy Ho:YLF pulsed lasers with good beam quality, $AgGaSe_2$ may find increasing applications for tunable IR generation down to 18 μm.

As a summary, Fig. 4.22 shows some typical schemes of OPG/OPA capable of producing picosecond tunable output from the UV (around 210 nm) to mid-IR (18 μm) using various commercial available crystals.

5. GENERATION OF TUNABLE FEMTOSECOND IR RADIATION VIA OPG/OPA

Ultrashort infrared pulses with a pulse duration around 100-femtosecond are very useful for spectroscopic and dynamic investigations. Widely tunable femtosecond pulses generated from OPO have been reviewed by Tang and Cheng.[19] While they must have a high repetition rate (10^8 Hz), the output energy per pulse is low (few nJ).[57,58]

Femtosecond pulses with an output energy in the order of 10^1–10^2 μJ per pulse and a tuning range from near IR to mid-IR can be generated by an OPG/OPA system. With a high energy fs pulsed laser system as the pump, namely an amplified CPM dye laser or a regeneratively amplified Ti:Sapphire laser with an output energy of few mJ per pulse, it is possible to have a femtosecond OPG/OPA system with a conversion efficiency around 10–25% and a tuning range limited only by absorption and phase-matching of the nonlinear crystal.

In a femtosecond OPG/OPA system, the group velocity mismatch (GVM) between the pump, signal and idler pulses in the nonlinear crystal is most important. As discussed in Section 2.2.3, it limits the effective interaction length of the optical parametric process in the crystal. For example, the effective interaction length in BBO is about 5 mm for 100-fs pump and signal pulse in the red and near IR, respectively. When the crystal is pumped by 100-fs pulses at 400 nm, the effective interaction length is only about 0.3 mm. Fortunately, the laser damage threshold of a crystal is much higher with femtosecond pulses. For BBO, it is more than 100 GW/cm^2 with 100-fs pulses in the red or near IR. A quick estimate would readily show that a 1-mJ, 100-fs pump pulse in the red, focused to 0.1 cm^2, yields a larger parametric gain in a 2-mm BBO crystal than the picosecond case discussed earlier. Thus, a femtosecond OPG/OPA system using BBO is certainly feasible. Similar estimates on other nonlinear crystals are also encouraging.

The first femtosecond "traveling wave" OPG/OPA system with an output energy of 100 μJ and a pulse duration of 150 fs was reported by Joosen et al.[59] BBO crystals employing a two-stage configuration similar to that of Fig. 3.2 and is shown in Fig. 5.1. The BBO crystals were cut at 20.3°, one (5-mm long) for OPG and the other (7-mm long) for OPA. The device was pumped by the output of an amplified CPM dye laser at 615 nm with an output energy of 0.6 mJ and a pulse duration of 80 fs. The pump beam was divided by a 30/70 beam splitter; the 30% part was used to pump the OPG and preamplifier stage, and the 70% part was used to pump the OPA. With the pump intensity at 50 GW/cm^2 in the OPG and 15 GW/cm^2 in the OPA, the signal output of the system was more than 100 μJ at 830 nm with a bandwidth of 34 nm and a pulse duration of 150 fs. The system is potentially tunable from 775 nm to 3 μm.

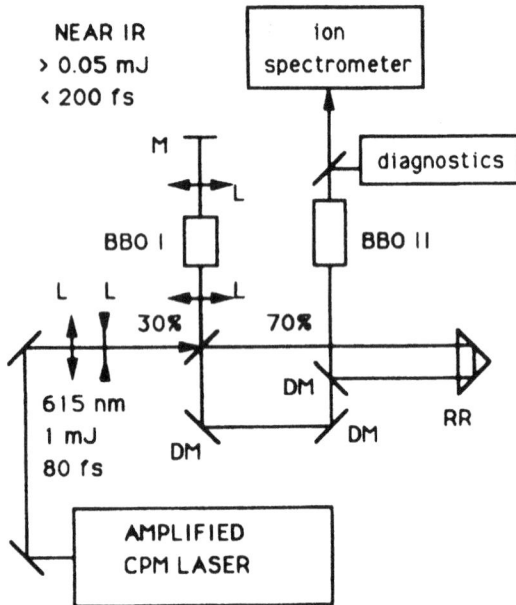

Fig. 5.1 The configuration of a two-stage femtosecond BBO-OPG/OPA system pumped by a CPM dye laser at 615 nm. [After W. Joosen *et al.*, *Opt. Lett.*, **27**, 133 (1992)].

More recently, Danielius *et al.*[60] and G. P. Banfi *et al.*[61] also constructed efficient fs-OPG/OPA systems pumped by a 1.3-ps and 190-fs pump pulses using a single-pass traveling wave configuration essentially the same as that shown in Fig. 3.1(b). The 1.3-ps OPG/OPA was achieved in BBO (with both type-I and type-II phase matching) and type-II KDP/type-I BBO crystals pumped by the frequency-doubled output of a passively mode-locked Nd:glass laser at 532 nm with a pulse duration of 1.3-ps and an output energy of 1 mJ. The 190-fs OPG/OPA was observed in BBO-type-I, BBO-type-II and LBO crystals pumped by 190-fs pulses generated by an amplified mode-locked dye laser with a pumping energy of 90 μJ at 600 nm. For the 1.3-ps pumped OPG/OPA systems, the maximum conversion efficiencies were about 25% for BBO and 13% for KDP/BBO (see Table 1). The tuning range was from 0.63 to 3.2 μm for the BBO-system and 0.77–1.7 μm for KDP/BBO-system. For the 190-fs-pumped OPG/OPA systems (see Table 2), the conversion efficiencies were around 20–25% and the tuning range was 0.75–3.1 μm. The bandwidth of the output was nearly transform-limited, Fig. 5.2 shows the conversion efficiency and spectral bandwidth versus signal wavelength of a type-II BBO-OPG/OPA system pumped by

190-fs pulses at the wavelength of 0.6 μm with an energy of 90 μJ per pulse. The BBO crystals used in the experiment were 0.8-cm long, while KDP used was 4-cm long. The pump intensities used in the 1.3-ps pumped OPG/OPA were 15 GW/cm^2 in type-I BBO, 20 GW/cm^2 in type-II BBO and 20 GW/cm^2 in KDP respectively, while in the 190-fs pumped BBO OPG/OPA system the pump intensities were 70 GW/cm^2 in OPG and 20 GW/cm^2 in OPA respectively.

A temperature-tuned LBO has a very small walk-off angle and group-velocity mismatch. It is therefore more suitable for fs OPG/OPA. Indeed it has been shown that a fs LBO-OPG/OPA system, using 15-mm LBO crystals with pump intensities of 25 GW/cm^2 in OPG and 8 GW/cm^2 in OPA, can be operated with a 200-fs pump pulse at an energy as low as 30 μJ. The output is tunable from 0.85 to 0.97 μm (signal) and from 1.6 to 2.1 μm (idler) when the temperature is changed from $\sim$20°C to 100°C. The maximum conversion efficiency of 15% occurs around 0.9 μm.[61] The temperature tuning curve of the system and its conversion efficiency and spectral width versus output signal wavelength are shown in Fig. 5.3. The characteristics of the 1.3-ps and 190-fs OPG/OPA systems discussed above are summarized in Tables 1 and 2.[60]

Table 1 Summary of the results of femtosecond OPG/OPA obtained with the 1.3-ps pump at 0.53 μm, E = 1 mJ, using a two-crystal (2-OPG) configuration.

OPG-stage	OPA-stage	$\Delta\nu_s$(cm^{-1})	η_{max}(%)	Tuning Range (μm)
BBO-type-I	BBO-type-I	200–1000	25	0.63–3.2
BBO-type-II	BBO-type-II	55@0.96 μm	20@0.9 μm	0.63–3.2
KDP-type-II	BBO-type-I	23–28	13	0.77–1.7

Table 2 Summary of the results of femtosecond OPG/OPA obtained with the 190-s pump at 0.6 μm using a three crystal (2-OPG + 1-OPA) configuration.

OPG	OPA	τ_s(fs)	$\Delta\nu_s$(cm^{-1})[a]		η_{max}(%)	Tuning Range (μm)
BBO-I	BBO-I	—	180	>1000	20–25[b]	0.75–3.1
BBO-I	BBO-II	180–250	120	220	23[b]	0.75–3.1
BBO-II	BBO-I	200–250	130	—	22–25[b]	0.75–3.1
LBO	LBO	200	200	300	15[c]	0.85–0.97
						(1.6–2.1)

[a]Minimum and maximum values are given for each crystal. [b]Pump energy = 90 microjoule. [c]Pump energy = 30 microjoule.

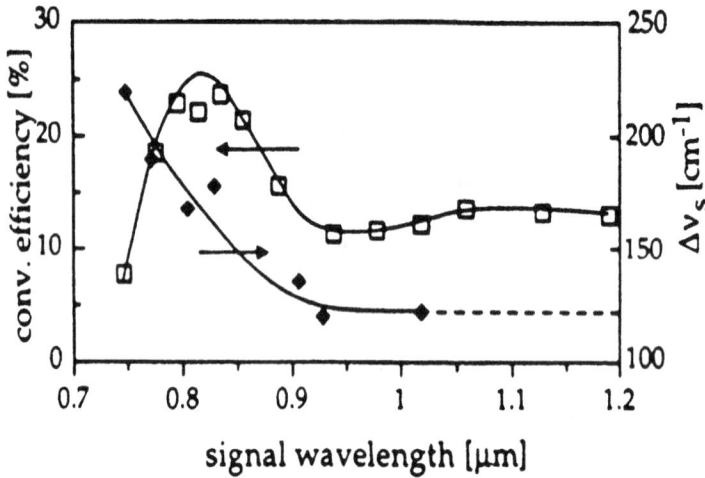

Fig. 5.2 Conversion efficiency and spectral bandwidths with type-II BBO OPG/OPA system pumped by an amplified mode-locked dye laser. The pump parameters are pulse length = 190 fs, pump wavelength = 0.6 μm and pump energy = 90 μJ. □ = efficiency and ♦ = bandwidth. The curves are only a guide for the eye. [After R. Danielius *et al.*, *J. Opt. Soc. Am.*, **B10**, 2222 (1993)].

The mode-locked Ti:sapphire laser with a regenerative amplifier is capable of producing 0.5–1 mJ/pulse output around 800 nm operating at a repetition rate of 1–10 KHz with a pulse duration around 100-fs. This system is ideally suitable to pump femtosecond OPG/OPA, as demonstrated by Zhang *et al.*[62] The OPG/OPA system was composed of two 6-mm BBO crystals cut for type-I phase matching and arranged in a double-pass configuration similar to that shown in Fig. 5.1. The Ti:sapphire laser operating at the wavelength of 780–840 nm, a pulse energy of 0.5 mJ, and a repetition rate of 1 KHz, was used to pump the OPG/OPA. The output with a tuning range from 1.2 to 2.5 μm is very stable and has a maximum conversion efficiency more than 10%. The detailed characteristics of the system will be published elsewhere. Two other groups have also reported successful operation of fs-OPA pumped by a regeneratively amplified mode-locked Ti:sapphire laser. Seifert *et al.*[63] used a 4-mm type-I BBO in a single stage, double-pass configuration and generated tunable IR in the range of 1.2–1.3 μm (signal branch) and 2.1–2.45 μm (idler branch). With a pump energy 200 μJ per pulse, the maximum OPA output is 5 μJ. Petrov *et al.*[64] used a 5-mm long type-II temperature tuned LBO in OPG stage and an 4-mm long type-I BBO in the amplifier using a traveling wave configuration. With a pump energy of 270 μJ, the generated OPA is tunable from 1.2 to 2.4 μm with a maximum output of 5 μJ,

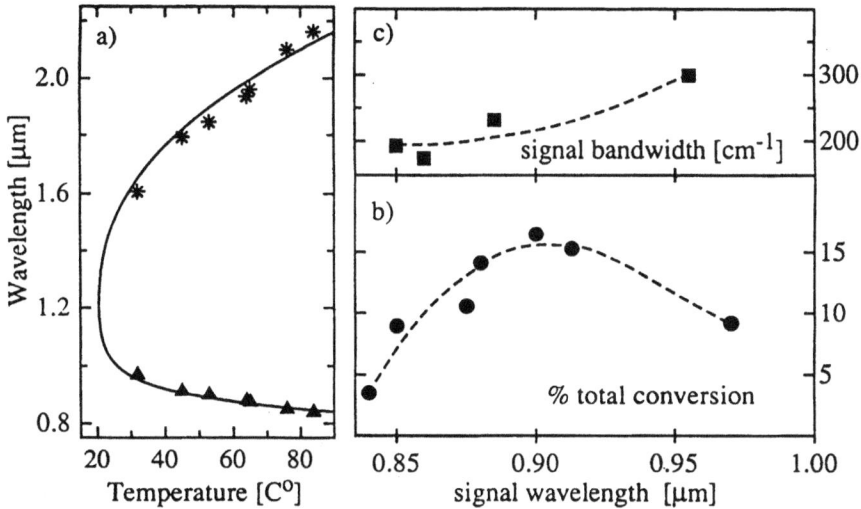

Fig. 5.3 (a) Temperature-tuning curve, (b) total conversion efficiency and (c) FWHM spectral width of the signal pulse of a femtosecond LBO-OPG/OPA system with NCPM. The pump parameters are = 190 fs, pump wavelength = 0.6 μm, E = 30 μJ. The curve in (c) is only a guide to the eye. [After G.P. Banfi *et al.*, *Opt. Lett.* **18**, 1633 (1993)].

corresponding to a total efficiency (signal + idler) of less than 5%. Both devices show a significant pulse shortening: with a pump pulse duration of 150–200 fs, Seifert *et al.*'s single-stage system has an output pulse duration of 70 fs in the generated OPA pulses, while Petrov *et al.*'s two-stage system produces pulses as short as 50 fs near 1460 nm. The pulse shortening was attributed to the broadband parametric amplification in the BBO amplifier.

6. SUMMARY

We have reviewed the OPG/OPA systems using various crystals pumped by pulsed lasers at various frequencies with pulse duration ranging from picoseconds to femtoseconds. As a summary, Table 3 provides a list of practical OPG/OPA systems and their characteristics.

Table 3 Summary of practical ultrashort-pulsed OPG/OPA systems.

Type	Pump Source			Specifications of OPG/OPA					Ref.
Material	λ_p (μm)	E_p (mJ)	τ_p (ps)	Tuning range (μm)	E_{out} (μJ)	$\Delta\nu$ (cm^{-1})	τ_{output} (ps)	η_{max} (%)	
BBO-I	0.355	3	15	0.42–2.6	<700	4–10	<10	30	[29]
LBO-I	0.355	2.3	15	0.42–2.3	<300	>10	—	—	[32]
	0.355	8	18	0.41–2.85	<900	7–20	14	28	[43][55]
KDP	0.355	1.35	35	0.45–1.6	<300	10	8.37	10	[49]
BBO	0.532	8	25	0.67–2.58	<520	>20	18	20	[44]
	0.532	5	25	0.63–2.6	<250	1–2	25	10	[32]
LBO	0.532	5	25	0.65–2.5	<450	4–10	18	24	[47]
KDP	0.532	—	30	—	—	40	—	50	[48]
KTP	0.526	12	32	0.64–4.0	<1500	26	—	20	[46]
LiNbO$_3$	0.532	0.7	7	0.59–3.7	—	20	—	10	[31]
LiNbO$_3$	1.064	1	6	1.4–4.0	<10	6.5	3.5	1	[41]
KTP	1.053	28	50	1.57–4.0	<1800	31	—	20	[46]
AgGaS$_2$	1.064	10	20	1.2–10	<100	8	10	10	[45]
BBO-I	0.532	1	1.3	0.63–3.2		200	>0.7	25	[60]
BBO-II	0.532	1	1.3	0.63–3.2		55	>0.7	20	[60]
KDP-II	0.532	1	1.3	0.77–1.7		23–28	>0.7	13	[60]
BBO	0.60	0.09	190	0.75–3.1		180	>180 fs	25	[60]
LBO	0.60	0.09	190	0.85–2.1		200	>200 fs	15@ E_p=30 μJ	[61]
BBO-I	0.78–0.84	0.5	100	1.2–2.5	<50		>100 fs	>10	[62]

7. CONCLUDING REMARKS

In recent years we have witnessed a strong revival of interest in optical parametric devices resulting from the technical advances in both the quality of the nonlinear crystals and the quality of the pump lasers. The wide range of tunability, the high output power, and the convenience associated with an all-solid-state device make them particularly attractive. It is now without question that such devices will largely replace dye lasers in the future as the preferred tunable radiation sources. Both optical parametric oscillators (OPO) and optical parametric amplifiers (OPA) have already become commercially available. Pulsed systems have obviously been the focus of the development. Output pulses can have pulsewidths ranging from nanosecond to picosecond and femtosecond, in both low and high repetition-rate modes. Output linewidths can be transform-limited, and output power can easily reach several hundred micojoules per pulse for the low rep-rate cases. Urged by the current overwhelming commercial interest in such devices, further development of OPO and OPA can be anticipated, presumably with more emphasis on reliability and output stability. It is our hope that this volume, together with the concurrent volume by C.L. Tang and L.K. Cheng on OPO to be published in this book series, will make a timely contribution to the future advance of this important technological area.

8. ACKNOWLEDGMENTS

This work was partially supported by National Science Foundation Grant No. PHY-9206769, the Faculty Research Committee of Georgia Southern University, by the Director, Office of Energy Research, Office of Basic Energy Sciences, Material Science Division of The US Department of Energy under Contract No. DE-AC03-765F00098, and by E.I. Dupont de Nemours & Co.

9. REFERENCES

1. See for example, the special issue *Optical Parametric Oscillation and Amplification*, eds. R. Byer and A. Piskarskas, *J. Opt. Soc. Am.* **B10**, No. 9 (1993).
2. *Conference on Nonlinear Optical Properties of Organic Materials IV, SPIE Proc.*, **1560** (1990).
3. *Conference on Physical Concepts of Materials for Novel Optoelectronic Device Application, SPIE Proc.*, **1361–2** (1990).
4. See for example, J. Hermann and B. Wilgelmi, *Lasers for Ultrashort Light Pulses* (Springer-Verlag, Berlin, 1987).
5. F. Krausz, M.E. Fermann, T. Brabec, P.F. Curley, M. Hofer, M.H. Ober, C. Spielmann, E. Wintner, and A.J. Schmidt, *IEEE J. Quant. Electron.* **QE-28**, 2097 (1992).
6. See for example, *Nonlinear Infrared Generation*, Y.R. Shen, Ed. (Springer-Verlag, Berlin, 1977); F. Zernike, and J.E. Midwinter, *Applied Nonlinear Optics* (Wiley, New York, 1973).
7. Y.R. Shen, *The Principles of Nonlinear Optics* (Wiley, New York, 1984).
8. G. Sciurba, and H.J. Loesch, *Opt. Comm.* **15**, 489 (1989).
9. A.L. Harris, and N.J. Levinos, *Appl. Opt.* **26**, 3996 (1987).
10. See for example, R.C. Eckardt, C.D. Nabors, W.J. Kozlovsky, and R.L. Byer, *J. Opt. Soc. Am.* **B8**, 646 (1991); R. Byer, Optical Parametric Oscillations, in *Treatise in Quantum Electronics*, Vol. I, Part B, H. Rabin, and C.L. Tang, eds. (Academic Press, New York, 1975), pp. 587.
11. J.A. Armstrong, N. Bloembergen, J. Ducuing, and P.S. Pershan, *Phys. Rev.* **127**, 1918 (1962).
12. J.A. Giordmaine, and R.C. Miller, *Phys. Rev. Lett.* **14**, 973 (1965).
13. C.C. Wang, and G.W. Racette, *Appl. Phys. Lett.* **6**, 169 (1965).
14. G.A. Boyd, and A. Ashkin, *Phys. Rev.* **146**, 189 (1966).
15. R.A. Baumgartner, and R.L. Byer, *IEEE J. Quant. Electron.* **QE-15**, 432 (1979).
16. D.E. Spence, P.N. Kean, and W. Sibbett, *Opt. Lett.* **42** (1991).
17. V.G. Dmitriev, G.G. Gurzadyan, and D.N. Nikogosyan, *Handbook of Nonlinear Optical Crystals* (Springer-Verlag, Berlin, 1991).
18. S.J. Brosnan, and R.L. Byer, *IEEE J. Quant. Electron.* **QE-15**, 415 (1979).
19. C.L. Tang, and L.K. Cheng, *Optical Parametric Processes and Oscillators* (Harwood Academic Publishers, forthcoming).
20. A.G. Akmonov, S.A. Akhmonov, R.V. Khokhlov, A.I. Kovrigin, A.S. Piskarskas, and A.P. Sukho-rukov, *IEEE J. Quant. Electron.* **QE-15**, 828 (1968); T.A, Rabson, H.J. Ruiz, P.L. Shah, and F.K. Tittle, *Appl. Phys. Lett.* **21**, 129 (1972); A. Laubereau, L. Greiter, and W. Kaiser, *Appl. Phys. Lett.* **25**, 87 (1974); A.H. Kung, *Appl. Phys. Lett.* **25**, 653 (1974), T. Kusgida, Y. Tanaka, M. Ojima, and Y. Nakazaki, *Jpn. J. Appl. Phys.* **14**, 1079 (1975); G.A. Dikchyus, V.I. Kabelka, A.S. Piskarskas, and A.Yu. Stabinis, *Sov. J. Quant. Electron.* **QE-4**, 1402 (1975); R. Danelyus, G. Dikchyus, V. Kabelka, A. Piskarskas, A.Yu. Stabinis, and Ya. Yasevichyute, *Sov. J. Ouant. Electron.* **QE-7**, 1360 (1977); W. Kranitzky, K. Ding, A. Seilmeier, and W. Kaiser, *Opt. Comm.* **34**, 483 (1980); F. Wondrazek, A. Seilmeier, and W. Kaiser, *Appl. Phys.* **B32**, 39 (1983); T. Elsaesser, A. Seilmeier, and W. Kaiser, *Opt. Comm.* **44**, 293 (1987); T. Elsaesser, H. Lobentanzer, and A. Seilmeier, *Opt. Comm.* **52**, 355 (1985); D.W. Anthon, H. Nathel, D.M. Guthals, and J.H. Clard, *Rev. Sci. Instrum.* **58**, 2054 (1987); U. Sukowski, and A. Seilmeier, *Appl. Phys.* **B50**, 541 (1990); H. Vangerzeele, *Appl. Opt.* **29**, 2246 (1990).
21. P.N. Butcher, and D. Cotter, *The Elements of Nonlinear Optics* (Cambridge University Press, Cambridge, 1990) chap. 2.
22. S.A. Akhmanov, and R.V. Khokhlov, *Nonlinear Optics* (Gordon and Beach, New York, 1972).
23. See ref. 17, chap. 2.

24. S.C. Sheng, and A.E. Siegman, *Phys. Rev.* **A21**, 599 (1980).
25. G.D. Boyd, and D.A. Kleinman, *J. Appl. Phys.* **39**, 3597 (1968).
26. R.A. Fisher, and W.K. Bischer, *Appl. Phys. Lett.* **23**, 661 (1973); *J. Appl. Phys.* **46**, 4291 (1975).
27. T.R. Taha, and M.J. Abloeitz, *J. Comp. Phys.* **55**, 201 (1984); S.M. Kopylov, *Sov. J. Quant. Electron.* **QE 11**, 920 (1981); Yu.N. Karamzin, and Zh. Vychisl, *Mat. Mat. Fiz.* **15**, 429 (1975).
28. J.Y. Huang, Y.R. Shen, C. Chen, and B. Wu, *Appl. Phys. Lett.* **58**, 1579 (1991).
29. J.Y. Huang, J.Y. Zhang, Y.R. Shen, C. Chen, and B. Wu, *Appl. Phys. Lett.* **57**, 1961 (1990).
30. J.Y. Zhang, J.Y. Huang, Y.R. Shen, C.Chen, and B.Wu, *Appl. Phys. Lett.* **58**, 213 (1991).
31. A. Seilmeier, and W. Kaiser, *Appl. Phys.* **23**, 113 (1980).
32. J.Y. Zhang, J.Y. Huang, Y.R. Shen, and C. Chen, *J. Opt. Soc. Am.* **B10**, 1758 (1993).
33. H. Zhou, J.Y. Zhang, T. Chen, C. Chen, and Y.R. Shen, *Appl. Phys. Lett.* **62**, 1457 (1993).
34. C. Chen, B. Wu, G. You, and Y. Huang, in *Digest of International Conference on Quantum Electronics* (Optical Society of America, Washington, D.C., 1984), paper MCC5; C. Chen, B. Wu, A. Jiang, and G. You, *Sci. Sinica Ser.* **28**, 235 (1985).
35. C.T. Chen, Y.C. Wu, A.D. Jiang, B.C. Wu, G.M. You, K.K. Li, and S.J. Lin, *J. Opt. Soc. Am.* **B6**, 616 (1989); S. Lin, Z. Sun, B. Wu, and C. Chen, *J. Appl. Phys.* **67**, 634 (1989); C. Chen, *Laser Focus World* **25**, 129 (Nov. 1989).
36. F.C. Zumsteg, J.D. Bierlein, and T.E. Gier, *J. Appl. Phys.* **47**, 4980 (1976); J.Q. Yao and T.S. Fahlen, *J. Appl. Phys.* **55**, 65 (1984); T.Y. Pan. C.E. Huang, B.Q. Hu, R.C. Eckardt, Y.X. Fan, R.L. Byer, and R.S. Feigelson, *Appl. Optics* **26**, 2390 (1987); R.C. Eckardt, H. Masuda, Y.X. Fan, and R.L. Byer, *IEEE J. Quant. Electron*, **QE-26**, 922 (1990).
37. M.M. Choy, and R.L. Byer, *Phys. Rev.* **B14**, 1693 (1976).
38. D.A. Bryan, R. Gerson, and H.E. Tomaschke, *Appl. Phys. Lett.* **44** (1984) 847; J.L. Nightgale, W.J. Silva, G.E. Reade, and R.L. Rybicki, *SPIE* **681**, 20 (1986).
39. R.A. Baumgartner, and R.L. Byer, *IEEE J. Quant. Electron.* **QE-15**, 432 (1979); R.C. Eckardt, Y.X. Fan, R.L. Byer, C.L. Marquardt, M.E. Storm, and L. Esterowitz, *Appl. Phys. Lett.* **49**, 608 (1986).
40. A. Laubereau, L. Greiter, and W. Kaiser, *Appl. Phys. Lett.* **25**, 87 (1974).
41. A. Seilmeier, K. Spanner, A. Laubereau, and W. Kaiser, *Opt. Comm.* **24**, 237 (1978).
42. W. Kranitzky, K. Ding, A. Seilmeier, and W. Kaiser, *Opt. Comm.* **34**, 483 (1980).
43. H.-J. Krause, and W. Daum, *Appl. Phys. Lett.* **60**, 2180 (1992).
44. X.D. Zhu, and L. Deng, *Appl. Phys. Lett.* **61**, 1490 (1992).
45. T. Elsaesser, A. Seilmeier, and W. Kaiser, *Appl. Phys. Lett.* **44**, 383 (1984).
46. H. Vanherzeele, J.D. Bierlein, and F.C. Zumsteg. *Technical Digest* from Conference on Nonlinear Properties of Materials, Troy, New York, 1988; H. Vanherzeele, *Appl. Opt.* **29**, 2246 (1990).
47. S. Lin, J. Huang, J. Ling, C. Chen, and Y.R. Shen, *Appl. Phys. Lett.* **59**, 2805 (1991).
48. V. Kabelka, A. Kutka, A. Piskarskas, V. Smil'gyavichyus, and Ya. Yasevichyute, *Sov. J. Quant. Electron.* **9**, 1022 (1979).
49. D.W. Anthon, H. Nathel, D.M. Guthals, and J.H. Clark, *Rev. Sci. Instrum.* **58**, 2054 (1987).
50. Y. Yang, Y. Cui, and M.H. Dunn, *Opt. Lett.* **17**, 192 (1992).
51. J.Y. Huang, Y.R. Shen, C. Chen, and B. Wu, *Appl. Phys. Lett.* **58**, 1579 (1991).
52. Y.X. Fan, R.C. Eckardt, R.L. Byer, R.K. Route, and R.S. Feigelson, *Appl. Phys. Lett.* **45**, 313 (1984).
53. T. Elsaesser, A. Seilmeier, and W. Kaiser, *Opt. Comm.* **44**, 29 (1983).
54. R.J. Seymour, and F. Zernike, *Appl. Phys. Lett.* **29**, 705 (1976); T. Elsaesser, H.J. Polland, A. Seilmeier, and W. Kaiser, *IEEE J. Quant. Electron.* **QE-20**, 191 (1984).
55. H.-J. Krause, and W. Daum, *Applied Physics*, **B56**, 8 (1993).
56. T. Dahinten, U. Plodereder, A. Seilmeier, K.L. Vodopyannov, K.R. Allakhverdiev, and Z.A. Ibragimov, *IEEE, J. Quant. Electron.* **QE-29**, 2245 (1993), and the references therein.
57. R. Danelyus, A. Piskarskas, and V. Sirutkaitis, *Sov. J. Quant. Electron.* **12**, 1626 (1982).

58. D.C. Edelstein, E.S. Wachman, and C.L. Tang, *Appl. Phys. Lett.* **54**, 1728 (1989); E.S. Wachman, D.C. Edelstein, and C.L. Tang, *Opt. Lett.* **15**, 136 (1990).

59. W. Joosen, H.J. Bakker, L.D. Noordam, H.G. Muller, and H.B. van Lidenvan Heuvell, *J. Opt. Soc. Am.* **B8**, 2087 (1991); W. Joosen, P. Agostini, G. Petite, J.P. Chambaret, and A. Antonetti, *Opt. Lett.* **27**, 133 (1992).

60. R. Danielius, A. Paskarskas, A. Stabinis, G.P. Banfi, P. Di Trapani, and R. Righini, *J. Opt. Soc. Am.* **B10**, 2222, (1993).

61. G.P. Banfi, R. Danielius, A. Piskarskas, D. Di Trapani, P. Foggi, and R. Righini, *Opt. Lett.* **18**, 1633 (1993).

62. J.Y. Zhang, Z. Xu, D. Deng, H. Wang, G.K. Wong, and K.S. Wong (forthcoming).

63. F. Seifert, V. Petrov, and F. Noack, *Opt. Lett.* **19**, 837 (1994).

64. V. Petrov, F. Seifert, and F. Noack, *Appl. Phys. Lett.* **65**, 268 (1994).

65. J.Y. Zhang, H.T. Zhou, and Y.R. Shen (unpublished experimental data).

Index

LASER SCIENCE AND TECHNOLOGY
An International Handbook

SECTIONS
Chaos and Laser Instabilities
Coherent Sources for VUV and Soft X-Ray Radiation
Distributed Feedback Lasers
Excimer Lasers
Fiber Optics Devices
Frequency Stable Lasers and Applications
Gas Lasers
Interaction of Laser Light with Matter
Laser Diagnostics in Chemistry
Laser Fusion
Laser Monitoring of the Atmosphere
Laser Photochemistry
Laser Spectral Analysis
Lasers and Communication
Lasers and Fundamental Physics
Lasers and Nuclear Physics
Lasers and Surfaces
Lasers in Medicine and Biology
Mechanical Action of Laser Light
New Solid State Lasers
Optical Bistability
Optical Computers
Optical Storage and Memory
Phase Conjugation
Semiconductor Diode Lasers
Solid State Lasers
Squeezed States of Light
Topics in Nonlinear Optics
Topics in Theoretical Quantum Optics
Tunable Lasers for Spectroscopy
Ultrashort Pulses and Applications
Vapour Deposition of Materials on Surfaces

PUBLISHED TITLES

For Product Safety Concerns and Information please contact our EU
representative GPSR@taylorandfrancis.com
Taylor & Francis Verlag GmbH, Kaufingerstraße 24, 80331 München, Germany

www.ingramcontent.com/pod-product-compliance
Lightning Source LLC
Chambersburg PA
CBHW061613220326
41598CB00024BC/3570